Photonics and Photoactive Materials

EUROMAT 2019: Symposium on Photonics and Photoactive Materials, September 1-5, 2019, Stockholm, Sweden.

Editor

Paolo Prosposito

Industrial Engineering Department and INSTM, University of Rome "Tor Vergata", Rome, Italy

Peer review statement

All papers published in this volume of "Materials Research Proceedings" have been peer reviewed. The process of peer review was initiated and overseen by the above proceedings editor(s). All reviews were conducted by expert referees in accordance to Materials Research Forum LLC high standards.

Published under License by **Materials Research Forum LLC**
Millersville, PA 17551, USA

Published as part of the proceedings series
Materials Research Proceedings
Volume 16 (2020)

ISSN 2474-3941 (Print)
ISSN 2474-395X (Online)

ISBN 978-1-64490-070-3 (Print)
ISBN 978-1-64490-071-0 (eBook)

Distributed worldwide by

Materials Research Forum LLC
105 Springdale Lane
Millersville, PA 17551
USA
http://www.mrforum.com

Manufactured in the United State of America
10 9 8 7 6 5 4 3 2 1

Table of Contents

Editorial

Committees

Characterization and Tests of Different Mach-Zehnder Silicon Photonic Modulator Configurations
Davide Badoni, Vincenzo Bonaiuto, Mauro Casalboni, Fabio De Matteis, Giovanni Di Giuseppe, Luca Frontini, Roberto Gunnella, Valentino Liberali, Andreas Mai, Giovanni Paoluzzi, Paolo Prosposito, Andrea Salamon, Gaetano Salina, Fausto Sargeni, Sigurd Schrader, Alberto Stabile, Patrick Steglich 1

Silver Nanoparticles with Different Thiol Functionalization: An Opposite Optical Behaviour in Presence of Hg(II)
Luca Burratti, Iole Venditti, Chiara Battocchio, S. Casciardi, Paolo Prosposito 6

Two Different Acid Oxidation Syntheses to Open C_{60} Fullerene for Heavy Metal detection
E. Ciotta, L. Burratti, P. Prosposito, E. Bolli, S. Kaciulis, S. Antonaroli, R. Pizzoferrato 16

Nonlinear Optical Characterization of $CsPbBr_3$ Nanocrystals as a Novel Material for the Integration into Electro-Optic Modulators
Francesco Vitale, Fabio De Matteis, Mauro Casalboni, Paolo Prosposito, Patrick Steglich, Viachaslau Ksianzou, Christian Breiler, Sigurd Schrader, Barbara Paci, Amanda Generosi 27

Plasma-Induced Generation of Optically Active Defects in Glasses
Christoph Gerhard 38

Preliminary Data on a SERS-Responsive Sensor Based on Metallic Nanostructures Functionalized by Aptamers Specific for Arsenic
Domenica Musumeci, Daniela Montesarchio, Elisa Scatena, Costantino Del Gaudio, Fabio De Matteis, Roberto Francini, Mauro Casalboni 46

Growth and Optical Characterisation of Lithium Fluoride Films for Proton Beam Detectors
Maria Aurora Vincenti, Mauro Leoncini, Stefano Libera, Enrico Nichelatti, Massimo Piccinini, Alessandro Ampollini, Luigi Picardi, Concetta Ronsivalle, Alessandro Rufoloni, Rosa Maria Montereali 56

Keyword Index

Editorial

This book collects some of the works presented at the A6 Symposium on Photonics and Photoactive Materials of the European Congress and Exhibition on Advanced Materials and Processes EUROMAT 2019 held in Stockholm – SW from 1 to 5 of September 2019. Recently, the development of materials and nanomaterials with specific funcionalization and the possible nanostructurization up to dimension small as never before, pushed the Photonics to an unbelievable level. Generation of light, absorption, emission, transmission, optical sensing and probing, signal processing and data transmission are only few of the properties related to this increasingly growing field. The topics treated in the book are related to fundamental properties and applications of innovative photonic and photoactive materials such as photonic structures, silicon photonics, nanomaterials, plasmonics, graphene quantum dots, optically active defects, fluorescent materials and optical sensors.

The book was published thanks to the financial support of Regione Lazio L.R. 13/08, Progetto Gruppo di Ricerca n. prot. 85-2017-15125.

Committees

Local Organizing Committee

Paolo Prosposito

University of Rome Tor Vergata, Italy

Hong Zhang

University of Amsterdam, The Netherlands

Xavier Le Guevel

University Grenoble-Alpes (UGA), France

Lining Sun

University of Shanghai, China

Sigurd Schrader

Technical University of Applied Sciences Wildau, Germany

Materials Research Proceedings **16** (2020) 1-5
https://doi.org/10.21741/9781644900710-1

Characterization and Tests of Different Mach-Zehnder Silicon Photonic Modulator Configurations

Davide Badoni[1,a*], Vincenzo Bonaiuto[2,b], Mauro Casalboni[2,c], Fabio De Matteis[2,d], Giovanni Di Giuseppe[3,e], Luca Frontini[4,f], Roberto Gunnella[3,g], Valentino Liberali[5,h], Andreas Mai[6,i], Giovanni Paoluzzi[1,j], Paolo Prosposito[2,k], Andrea Salamon[1,l], Gaetano Salina[1,m], Fausto Sargeni[2,n], Sigurd Schrader[7,o], Alberto Stabile[5,p], Patrick Steglich[6,7,q]

[1]INFN Structure of Rome Tor Vergata, Rome, Italy

[2]University of Rome Tor Vergata, Rome, Italy

[3]University of Camerino, Macerata, Italy

[4]INFN Structure of Milano, Milan, Italy

[5]University of Milano, Milan, Italy

[6]IHP – Innovations for High Performance Microelectronics, Frankfurt (Oder), Germany

[7]Technical University of Applied Sciences Wildau, Wildau, Germany

[a]davide.badoni@roma2.infn.it, [b]vincenzo.bonaiuto@uniroma2.it, [c]mauro.casalboni@roma2.infn.it, [d]fabio.dematteis@roma2.infn.it, [e]gianni.digiuseppe@unicam.it, [f]luca.frontini@mi.infn.it, [g]roberto.gunnella@unicam.it, [h]valentino.liberali@mi.infn.it, [i]andreas.mai@ihp-microelectronics.com, [j]giovanni.paoluzzi@roma2.infn.it, [k]paolo.prosposito@roma2.infn.it, [l]andrea.salamon@roma2.infn.it, [m]gaetano.salina@roma2.infn.it, [n]fausto.sargeni@uniroma2.it, [o]schrader@th-wildau.de, [p]alberto.stabile@mi.infn.it, [q]psteglich@th-wildau.de

* Corresponding author

Keywords: CMOS, Mach-Zehnder, VLSI

Abstract. We designed and produced an integrated silicon photonic circuit, in a single chip with IHP SG25H4_EPIC 0.25 µm technology. A Mach-Zehnder interferometer with an alternative shape for better integration, together with a standard-shape Mach-Zehnder interferometer have been realized. In this work, preliminary results of comparative performance measurements between the two Mach-Zehnder interferometer are shown.

Introduction

Silicon photonics is a rapidly emerging field in research and technology of photonic integrated circuits [1]. Given its high refraction index and low dispersion, silicon is a high quality material for light guiding devices. On the other hand, VLSI CMOS compatible processes allow one to design reliable photonic structures, which can be easily integrated with standard electronic circuits at affordable cost [2,3]. Several structures can be envisaged by means of Silicon Photonic technology [4] such as, for example, optical ring resonators [5,6] or Mach-Zehnder interferometers.

In the basic standard geometry of the Mach-Zehnder interferometer, the arm dimensions are too large for an efficient integration on a silicon chip, in applications where a large number of interferometers are required such as in silicon pixel detectors used in many fields of science and applications - from particle and nuclear physics experiments to medical physics.

Optical modulation, in silicon photonic integrated Mach-Zehnder interferometers can be obtained using the interference condition between the two different interferometer branches by controlling the electro-optical effect (plasma dispersion) [7,8].

The silicon photonic integrated Mach-Zehnder, if available with a geometry that maximizes the integration on a silicon chip, is an excellent device for many applications that require integration of a large number of channels in a single chip. That allows realizing a chip that is able, together with the photonic element, to host electronics and sensors on silicon, such as high-speed data transmission in the high-energy particles silicon detectors. [9]

Two Mach-Zehnder interferometers with different configurations have been designed and produced in a single die in IHP SG25H4_EPIC 0.25 µm technology: one developed almost entirely in line (A configuration), while the second has the two arms folded in a spiral to minimize the occupied surface (B configuration) as shown in Figure 1.

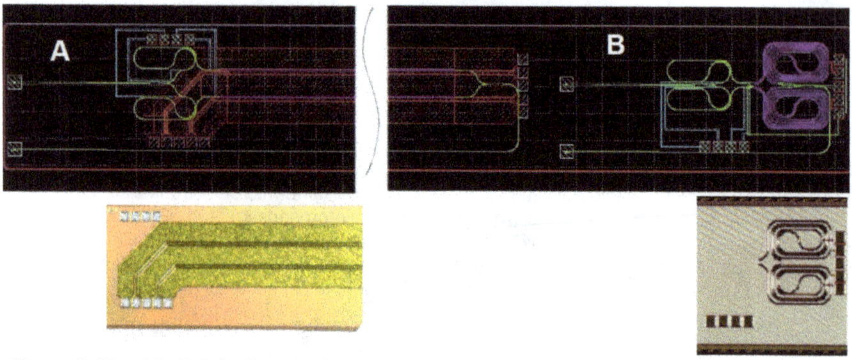

Figure 1: Two Mach-Zehnder interferometers with different geometric configuration in a single die. On the left is the standard straight configuration (A) while on the right there is a spiral configuration (B). On the top, the layout drawings are visible; on the bottom are pictures of details corresponding to devices.

The spiral configuration (1300 µm x 700 µm), with respect to the standard straight one (6200 µm x 600 µm), offers smaller size which allows easier integration.

Test setup

The two different Mach-Zehnder interferometers have been tested in the laboratory and their performances have been compared in a 1550 nm optical telecommunication window.

The optical and electrical contact positioning has been carried out by means of micrometric movements on three perpendicular axes.

Measurements and test results.

The first test consists in a transmission spectrum measurement around the typical wavelength window in the absence of electrical polarization (V = 0 Volt), as shown in Figure 2 and Figure 3.

Materials Research Forum LLC
https://doi.org/10.21741/9781644900710-1

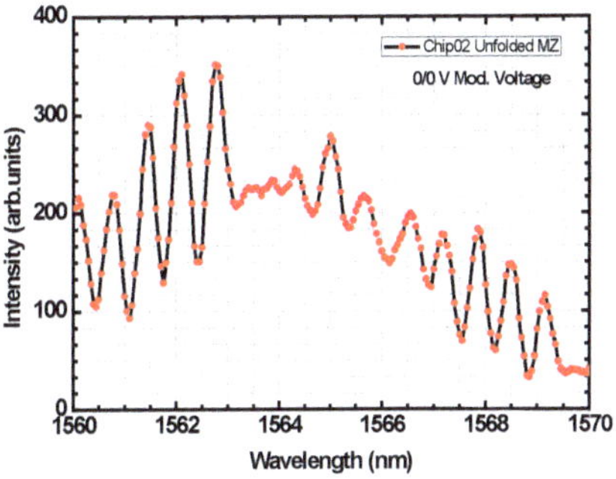

Figure 2: Unfolded Mach-Zehnder interferometer trasmission spectrum.

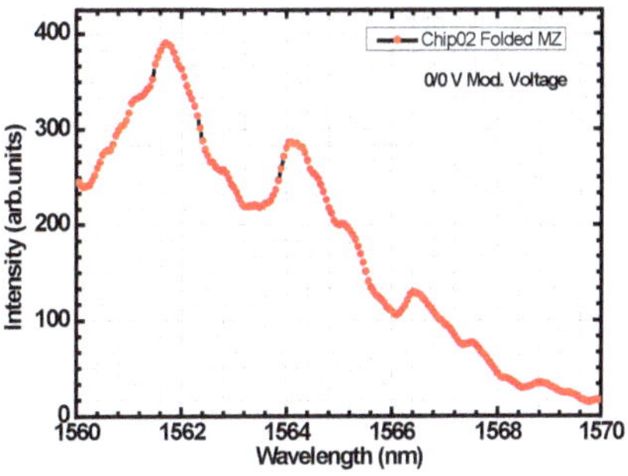

Figure 3: Folded Mach-Zehnder interferometer trasmission spectrum.

The oscillation of the transmitted intensity at different wavelengths is due to frequency dependent dephasing of the light transmitted along the two arms of the Mach-Zehnder. The second test consists in a modulation scan carried out at wavelength fixed by applying a reverse potential ramp (0V / 3V) to a single interferometer Mach-Zehnder arm as shown in Figure 4 and Figure 5.

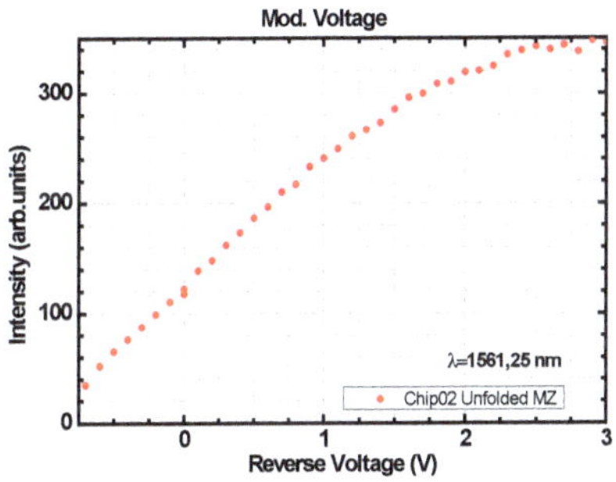

Figure 4: Unfolded Mach-Zehnder interferometer modulation scan vs. potential ramp.

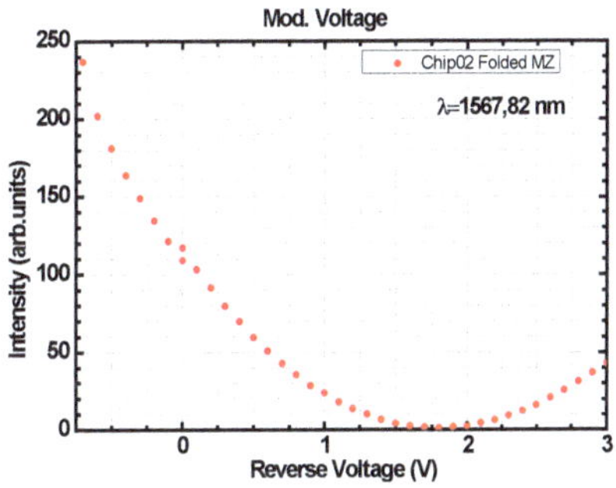

Figure 5: Folded Mach-Zehnder interferometer modulation scan vs. potential ramp.

In both cases, a large area of linear modulation of the signal is clearly shown as a function of the applied potential. The signal is coupled and transmitted with good efficiency in the optical circuit.

The last test performed was to observe the dynamic modulation response of the Mach-Zehnder interferometer. A laser source injected the beam in the Mach-Zehnder and the output of the modulator optically was coupled with an ultrafast, InGaAs PIN photodetector Picometrix P-50A. The signal was amplified by a SHF 806 E modulator drive. The modulation was obtained with a square wave applied to the modulator diodes. The results are shown in Figure 6, where the resulting modulation is clearly visible.

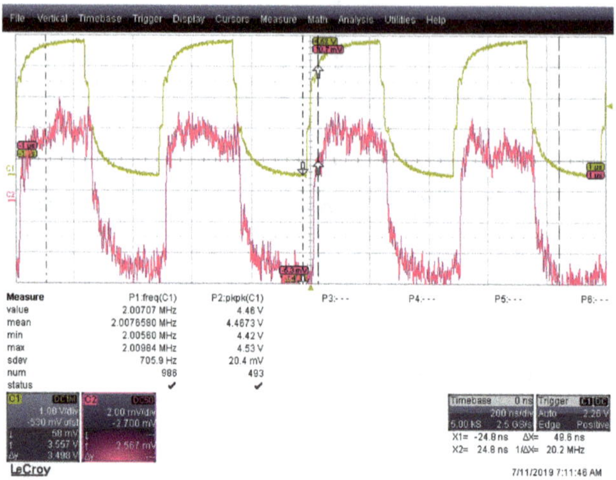

Figure 6: Acquisition from oscilloscope of the modulating electrical signal applied to the MZ junction and the resulting optical signal output from the MZ coupled with an InGaAs PIN photodetector Picometrix P-50A and amplified.

Conclusions and Outlook

An integrated silicon photonic chip implementing two different Mach-Zehnder interfermeters, in straight and folded configurations, was produced in in a single die in IHP SG25H4_EPIC 0.25 µm technology. Some preliminary tests were performed and both Mach-Zehnder interferometers showed clear electro-optical modulation. More detailed high speed bit error rate data transmission tests with standard telecom pseudo random binary sequences are foreseen to validate the design. The folded Mach-Zehnder interferometer, given its reduced footprint, represents a promising device in many space critical applications.

Acknowledgements

The authors want to thank Dr. Lisa Noetzel who reviewed the paper.

References

[1] P. Steglich and F. De Matteis, "Introductory Chapter: Fiber Optics" In Fiber optics Ed. P. Steglich, IntechOpen, (2019) ISBN: 978-1-83881-156-3. https://doi.org/10.5772/intechopen.74877

[2] R. Russo et al., Toward optical and superconducting circuit integration, Supercond. Sci. Tech., 17(5), S456-S459 (2004) 45. https://doi.org/10.1088/0953-2048/17/5/074

[3] P. Steglich et al., Hybrid-Waveguide Ring Resonator for Biochemical Sensing, IEEE Sensors J., 17(15), 4781-4790 (2017). https://doi.org/10.1109/JSEN.2017.2710318

[4] F. Bonaccorso et al, Graphene photonics and optoelectronics, Nat Photonics 4, 611–622 (2010). https://doi.org/56 10.1038/nphoton.2010.186

[5] G. Alimonti, et al., Use of silicon photonics wavelength multiplexing techniques for fast parallel readout in high energy physics, Nuclear Inst. and Methods in Physics Research: A, 936, 601 (2019). https://doi.org/10.1016/j.nima.2018.09.088

[6] P. Prosposito et al, UV-nanoimprinting lithography of Bragg Gratings on hybrid sol-gel based channel waveguides, Solid 47 State Sci. 12, 1886-1889 (2010). https://doi.org/10.1016/j.solidstatesciences.2010.03.014

[7] R. A. Soref and B.R. Bennett, Electrooptical effects in silicon, IEEE J Quantum Elect QE-23 (1), 123, (1987). https://doi.org/1 10.1109/JQE.1987.1073206

[8] G.T. Reed and E.J.Png, Silicon optical modulators, Mater.Today, 8(1), 40-50 (2005). https://doi.org/10.1016/S1369-7021(04)00678-9

[9] E. Cortina Gil et al., The beam and detector of the NA62 experiment at CERN, JINST 12 P05025 (2017).

Silver Nanoparticles with Different Thiol Functionalization: An Opposite Optical Behaviour in Presence of Hg(II)

Luca Burratti[1,a *], Iole Venditti[2,b], Chiara Battocchio[2,c], S. Casciardi[3,d], Paolo Prosposito[1,e]

[1] Department of Industrial Engineering and INSTM, University of Rome Tor Vergata, Via del Politecnico 1, Rome, 00133, Italy

[2] Department of Sciences, Roma Tre University of Rome, Via della Vasca Navale 79, 00146 Rome, Italy

[3] National Institute for Insurance against Accidents at Work (INAIL), Department of Occupational and Environmental Medicine, Epidemiology and Hygiene, 00078 Monte Porzio Catone, Italy

[a]luca.burratti@uniroma2.it, [b]iole.venditti@uniroma3.it, [c]chiara.battocchio@uniroma3.it, [d]s.casciardi@inail.it, [e]paolo.prosposito@uniroma2.it

Keywords : Metal Nanomaterials, Silver Nanoparticles, Localized Surface Plasmon Resonance (L-SPR), Optical Sensors, Hg(II) Ions Detection

Abstract. We synthesized two different functionalized silver nanoparticles (AgNPs) in water, starting from silver nitrate, as Ag(I) ions precursor, and sodium borohydride, as reduction agent. The first system was capped with sodium 3-mercapto-1- propansulfonate (3MPS), while L-Cysteine and citrate stabilized the other system. We characterized both systems by UV-Vis absorption spectroscopy and transmission electron microscopy (TEM). We tested their optical response to several heavy metal ions monitoring the Localized Surface Plasmon Resonance (LSPR) band. In particular, these two systems have an opposite optical behaviour in presence of Hg(II) ions as contaminants. In the case of AgNPs-L-Cysteine/citrate, the plasmonic band shifted to higher wavelengths affording a linear behaviour and LOD, in the range from 1 to 7.5 ppm and 600 ppb, respectively; whereas, the AgNPS-3MPS peak shifted to lower wavelengths with a linear range from 0 to 5 ppm and a LOD of 240 ppb for Hg(II). A preliminary hypothesis about the interaction mechanism between AgNPs and Hg(II) ions is discussed.

Introduction

Micro and nano-materials, display unique physical and chemical properties with respect to the same materials in bulk structure. These particular features are widely exploited in numerous fields, such as energy, optoelectronics, biomedicine and sensors[1–9]. In the field of sensors, optical properties can be based on reflectivity, photoluminescence emission and absorption of micro/nano-systems. When certain contaminants, such as small molecules, volatile organic compounds (VOCs), pesticides or heavy metal ions (HMIs) interact with the optical device, they are able to change the optical signal on which the sensor is based. In general, the variation of the signal can regard its intensity, its energy or its shape. These changes are related to the different nature of the interaction between the optical device and the external stimulus.

Among sensors based on light reflection, photonic crystals (PCs) show a Bragg-like diffraction in the visible range as a consequence of a peak in the reflectance spectrum[10–12]. The porous structure of PCs allows the permeation of contaminants between the spheres and a change of dielectric function of the whole device occurs causing a shift of the reflectivity band. Sensors based on PCs have been employed for the detection of several hazardous compound such us VOCs[13–16].

Although inorganic quantum dots (QDs), graphene oxide quantum dots (GOQDs) or metal nanoclusters (MNCs), have different chemical structures and are obtained with different synthetic strategies[17–22], they are linked by a common feature: the photoluminescence emission (PL). Also, in these cases, photoemitting material have been used to detect the presence of various species of pollutants[23–28]. The interaction mechanism would explain the opposite behaviour in the PL intensity change[29].

Among the sensors based on a change in optical absorption[30–32], metal nanoparticles (MNPs) show a typical absorption peak in the UV-Vis range, due to the phenomenon of Localized-Surface Plasmon Resonance (LSPR)[33]. To prevent the aggregation of these MNPs, during the chemical synthesis, a stabilizing molecule, which is also a species sensitive to the presence of the pollutants, is introduced. Colloidal solutions of MNPs show a specific colour depending on the type of noble metal (gold, silver or cupper), on the size and shape of the NPs[34], thus a shift of the plasmonic band produces a change of colour of the solution, which can be appreciated by naked eye (colorimetric sensors)[35–38]. The study and the development of optical sensors should be encouraged for the simplicity of the synthesis, the fast response to the external stimulus and the ease of data interpretation.

In this work, two different AgNPs systems functionalized with hydrophilic capping agents were synthesized, using citric acid (Cit) and L-Cysteine (L-Cys) for the first Ag-nanosystem and 3-mercapto-1-propanesulfonate (3MPS) for the second one. We characterized both systems, by Dynamic Light Scattering (DLS) and Transmission Electron Microscopy (TEM) analysis confirming nanosized dimensions. AgNPs- 3MPS and AgNPs-L-Cys/Cit were tested as plasmonic sensor for heavy metal detection in water, showing a good response for Hg(II) ions in both cases, but with an opposite behaviour. A hypothesis about the interaction mechanism between AgNPs and Hg(II) ions were discussed.

Experimental section Chemical reagents

Silver nitrate ($AgNO_3$), Sodium borohydride ($NaBH_4$), Sodium 3-mercapto-1-propanesulfonate (3MPS), Sodium citrate ($Na_3C_6H_5O_7$, Cit), L-Cysteine ($C_3H_7NO_2S$, L-Cys) have been purchased from Sigma-Aldrich. For the sensing tests we used the following metallic salt: $NaAsO_2$, $NaHAsO_4 * 7H_2O$, $Ca(ClO_4)_2$, $Cd(NO_3)_2$, $CoCl_2 * 6H_2O$, $CrCl_3 * 6H_2O$, $Cu(NO_3)_2$, $FeCl_3 * 6H_2O$, $Hg(NO_3)_2 * H_2O$, $KClO_4$, $Mg(ClO_4)_2$,

$NaClO_4$, $NdCl_3 * 6H_2O$, $NiCl_2 * 6H_2O$, $Pb(NO_3)_2$, $Zn(NO_3)_2 * 6H_2O$ purchased from Sigma-Aldrich. All chemical reagents have been used as received without any further process and all solutions have been prepared with deionized water (resistivity 18.2MΩcm at 25°C) obtained from Millipore Milli-Q water purification system.

Apparatus

Optical characterizations of bare AgNPs systems and contamination tests have been performed by Perkin-Elmer Lambda 19 UV/Vis/NIR spectrophotometer, in the range from 300 nm to 700 nm. The morphological characterization of the AgNPs has been accomplished with a Transmission Electron Microscope (TEM). The experimental apparatus is a FEI TECNAI 12 G2 (120 KeV) equipped with an energy filter (GATAN GIF model) and a Peltier cooled SSC (slow scan charged coupled device) multiscan camera (794 IF model).

Materials Research Proceedings **16** (2020) 6-15
https://doi.org/10.21741/9781644900710-2

Synthesis of AgNPs

The AgNPs stabilized by 3MPS were prepared by a wet reduction of $AgNO_3$ with $NaBH_4$. The detailed synthesis is reported in our previous works [39,40]. Briefly, the $AgNO_3$ water solution is added dropwise to $NaBH_4$ solution under vigorous stirring at the temperature of 3°C; 3MPS was subsequently added and it capped the silver nanoparticles. After the synthesis is completed, the solution is stored at T = 4°C before the analysis.

The AgNPs stabilized with citrate and L-Cysteine were prepared and characterized in analogy to literature reports [41–43]. Sodium citrate, L-Cys and $AgNO_3$ were dissolved separately in distilled water and then the three solutions were mixed under magnetic stirring. After degassing the mixture with Argon for 10 minutes, a $NaBH_4$ solution were added and allowed to react at room temperature for 2 hours. The obtained brown solution was purified by centrifugation, collecting the AgNPs in the precipitate and resuspended in water (13000 rpm, 10 min, 2 times with deionized water).

Results and discussion

Fig.1 shows a TEM image of the AgNPs-3MPS system: the particles have a mean diameter of about 4 nm. The dimensional characterizations obtained by TEM analysis for both systems are listed in Table 1

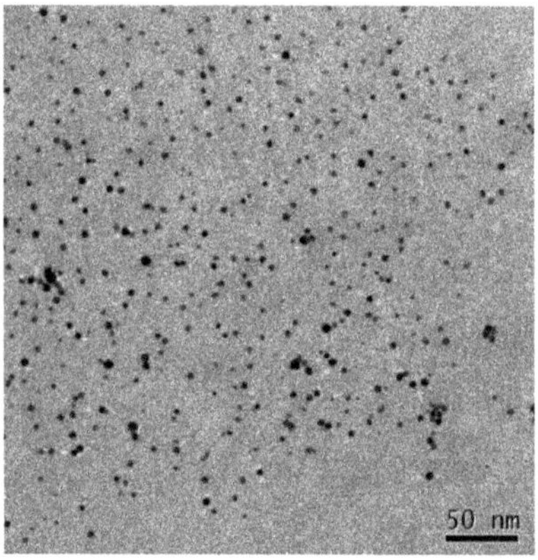

Fig. 1. *TEM image of the AgNPs-3MPS system.*

Fig. 2 shows the optical absorption in the UV-Vis range for both AgNPs systems. The solid black line represents the AgNPs-3MPS solution; the plasmon band absorption has the maximum centred at 396 nm and a Full Width at Half Maximum (FWHM) of 74 nm. The green dashed curve is the absorption of the reference solution of AgNPs- L-Cys/Cit, in this case the maximum of the peak is at 401 nm and the FWHM is 102 nm.

Table 1. *TEM size comparison.*

System	Size by TEM	REF.
AgNPs-3MPS	4.1 ± 0.4 nm	[39]
AgNPs-L-Cys/Cit	5 ± 2 nm	[43]

When both systems were contaminated with 5ppm of Hg(II) ions showed an opposite behaviour, the LSPR band of AgNPs-3MPS system shifted from about 400 to 350 nm (blue solid line), whereas for the AgNPs-L-Cys/Cit from 400 to around 430 nm (red dashed curve). Measuring the effect of different Hg(II) concentrations with step of 1 ppm, we obtained a linear behaviour from 0 to 5 ppm and a limit of detection (LOD, 3σ) of 240 ppb for the AgNPs-3MPS system. Similarly, for the other silver nanoparticles system, we obtained a linear behaviour between 1 to 7.5 ppm and we estimated a LOD of 600 ppb. We tested the optical response of both systems in presence of several metal ions checking the shape and energy of the LSPR at the concentration of 5 ppm. For some of them we did not find any difference in the respective LSPRs, for some others we found a similar behaviour, namely a shift of the LSPR band to longer wavelengths (red shift). The optical behaviour of the surface plasmon resonance band for the two silver systems in presence of the different heavy metal ions at the concentration of 5 ppm is reported in Table 2. In the table, a positive value represents a red shift, while a negative value is referred to a blue shift. The AgNPs-3MPS system presented a low selectivity, showing a similar optical response towards Cd(II), Co(II), Cu(II), Mg(II) and Ni(II) (red shift), but evidencing in the case of Hg(II) an opposite behaviour with respect to the others contaminants (blue, more marked shift). Furthermore, the AgNPS-L-Cys/Cit was more selective with respect to the other system, since it responded with a red shift to Cr(III), Cu(II) and Hg(II) even if it is less sensitive. However, as already outlined, the opposite shift of the two systems in presence of mercury ions, makes this combined optical sensor really selective for that specific contaminant. A preliminary hypothesis about the different interaction mechanism for Hg(II) of the two systems, that can justify the opposite behaviour, is discussed in the next section.

A preliminary hypothesis of the opposite optical behaviour in presence of Hg(II)

Some preliminary considerations about the coverage level of the capping agents for the two AgNPs systems has to be done. The AgNPs-3MPS system has a capping agent amount that is about 10 times lower than the $AgNO_3$ content, which could mean the 3MPS molecules cover only partially the AgNPs surface. The low coverage level allows the Hg(II) ions to easily reach the NPs surface and to react with it, forming an amalgam with the external Ag atoms[44,45].

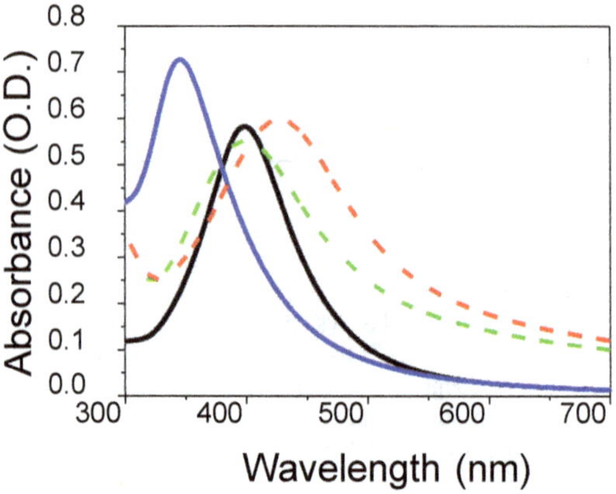

Fig. 2. *UV-Vis absorption spectra of both AgNPs systems without and with 5ppm of Hg(II). AgNPs-3MPS as reference and with 5ppm of Hg(II) contamination are the solid black and the solid blue curve, respectively. AgNPs-L-Cys/Cit as reference and with 5ppm of Hg(II) contamination are the dashed green and the dashed red curve, respectively.*

Table 2. *heavy metal ions tested with both systems.*

	Ag nano-system	
Metal ions at 5 ppm	NPs-3MPS	NPs-L-Cys/Cit
As(III)	\	\
As(V)	\	\
Ca(II)	\	\
Cd(II)	+ 11 nm	\
Co(II)	+ 17 nm	\
Cr(III)	\	+ 8 nm
Cu(II)	+ 6 nm	+ 6 nm
Fe(III)	\	\
Hg(II)	- 44 nm	+ 26 nm
K(I)	\	\
Mg(II)	+ 6 nm	\
Na(II)	\	\
Nd(III)	\	\
Ni(II)	+ 19 nm	\
Pb(II)	\	\
Zn(II)	\	\

The mechanism could be explained on the basis of the electrochemical differences of Ag(I) and Hg(II) ions. The standard reduction potential for Ag is +0.80V (Ag(I) + e^- = Ag(0)) whereas for Hg(II) it is +0.85V (Hg(II) + $2e^-$ = Hg(0)) and according to the electrochemical series, metals with a higher reduction potential act as better oxidising agents[46,47]. Therefore,

Hg(II) ions will reduce at the surface of AgNPs, while Ag atoms will oxidize, giving origin at the amalgam. This mechanism could explain the blue shift of LSPR band of AgNPs-3MPS.

In the other case, the two stabilizers (citrate and L-Cysteine) are about 14 times higher than the Ag(I) ions, thus the coverage of AgNPs surface is probably complete. The stability of AgNPs colloidal solutions is due to the electrostatic repulsions of functional groups of capping agents, indeed, L-Cys and Cit molecules with their carboxylate ($-COO^-$) groups prevent the aggregation of AgNPs in solution. When the AgNPs-L-Cys/Cit solution is in presence of Hg(II) ions, the latter cannot reach the bare surface of NPs. This favours the aggregation of the particles, and as consequence, we observe a red shift of the optical absorption and a broadening of the LSPR band.

Conclusions

In this study, we synthesized two different functionalized AgNPs in water, with a narrow size dispersion and a good stability over the time. We tested their optical response to several heavy metal ions monitoring the LSPR band before and after adding contaminated water. An opposite optical behaviour for the two systems in presence of Hg(II) ions have been measured. In the case of AgNPs-L- Cysteine/citrate, the plasmonic band shifted to higher wavelengths while in the case of AgNPS-3MPS shifted to lower wavelengths. This behaviour offers a very good selectivity in the detection of Hg(II) in water with a LOD of about 600 ppb. The opposite behaviour can be related to the different coverage level of NPs surface. For 3MPS Ag system, the capping agent amount is not enough to cover completely the NPs, thus the Hg(II) ions interact directly with the Ag external atoms, forming an amalgam which give rise to the blue shift in the absorption spectrum. For the other nano-system, where the coverage level is enough to completely shield the NPs surface, the Hg(II) ions interact with the particle surface, promoting the aggregation of AgNPs and as a consequence a red shift of the absorption spectrum was detected.

Acknowledgement

This research was partially funded by Regione Lazio, through Progetto di ricerca 85- 2017-15125, according to L.R.13/08 and by the University of Rome Tor Vergata in the framework of "GHOST" project, within "Mission Sustainability" program (D.R. 2817/2016), grant number (CUP): E86C18000450005. The Grant of Excellence Departments, MIUR (ARTICOLO 1, COMMI 314 – 337 LEGGE 232/2016), is gratefully acknowledged by authors of Roma Tre University.

References

[1] X. Cao, C. Tan, X. Zhang, W. Zhao, H. Zhang, Solution-Processed Two- Dimensional Metal Dichalcogenide-Based Nanomaterials for Energy Storage and Conversion, Adv. Mater. 28 (2016) 6167–6196. https://doi.org/10.1002/adma.201504833

[2] P. Prosposito, L. D'Amico, M. Casalboni, N. Motta, Periodic arrangement of mono-dispersed gold nanoparticles for high performance polymeric solar cells, in: 2015 IEEE 15th Int. Conf. Nanotechnol., IEEE, 2015: pp. 378–380. https://doi.org/10.1109/NANO.2015.7389005

[3] F. Xia, T. Mueller, Y. Lin, A. Valdes-Garcia, P. Avouris, Ultrafast graphene photodetector, Nat. Nanotechnol. 4 (2009) 839–843. https://doi.org/10.1038/nnano.2009.292

[4] S.I. Valyansky, E.K. Naimi, L. V. Kozhitov, Functional 2D nanomaterials for optoelectronics based on langmuir bacteriorhodopsin films, Mod. Electron. Mater. 2 (2016) 79–84. https://doi.org/10.1016/j.moem.2016.12.007

[5] S. Priyadarsini, S. Mohanty, S. Mukherjee, S. Basu, M. Mishra, Graphene and graphene oxide as nanomaterials for medicine and biology application, J. Nanostructure Chem. 8 (2018) 123–137. https://doi.org/10.1007/s40097-018-0265-6

[6] J.J. Giner-Casares, M. Henriksen-Lacey, M. Coronado-Puchau, L.M. Liz- Marzán, Inorganic nanoparticles for biomedicine: where materials scientists meet medical research, Mater. Today. 19 (2016) 19–28. https://doi.org/10.1016/j.mattod.2015.07.004

[7] M. Etienne, A. Goux, E. Sibottier, A. Walcarius, Oriented Mesoporous Organosilica Films on Electrode: A New Class of Nanomaterials for Sensing, J. Nanosci. Nanotechnol. 9 (2009) 2398–2406. https://doi.org/10.1166/jnn.2009.SE39

[8] P.K. Kannan, D.J. Late, H. Morgan, C.S. Rout, Recent developments in 2D layered inorganic nanomaterials for sensing, Nanoscale. 7 (2015) 13293–13312. https://doi.org/10.1039/C5NR03633J

[9] R. De Angelis, I. Venditti, I. Fratoddi, F. De Matteis, P. Prosposito, I. Cacciotti, L. D'Amico, F. Nanni, A. Yadav, M. Casalboni, M. V. Russo, From nanospheres to microribbons: Self-assembled Eosin Y doped PMMA nanoparticles as photonic crystals, J. Colloid Interface Sci. 414 (2014) 24–32. https://doi.org/10.1016/j.jcis.2013.09.045

[10] J. Hou, H. Zhang, Q. Yang, M. Li, L. Jiang, Y. Song, Hydrophilic- Hydrophobic Patterned Molecularly Imprinted Photonic Crystal Sensors for High-Sensitive Colorimetric Detection of Tetracycline, Small. 11 (2015) 2738–2742. https://doi.org/10.1002/smll.201403640

[11] Z. Cai, N.L. Smith, J.-T. Zhang, S.A. Asher, Two-Dimensional Photonic Crystal Chemical and Biomolecular Sensors, Anal. Chem. 87 (2015) 5013– 5025. https://doi.org/10.1021/ac504679n

[12] Y. Zhang, Y. Zhao, R. Lv, A review for optical sensors based on photonic crystal cavities, Sensors Actuators A Phys. 233 (2015) 374–389. https://doi.org/10.1016/j.sna.2015.07.025

[13] C. Dispenza, M.A. Sabatino, S. Alessi, G. Spadaro, L. D'Acquisto, R. Pernice, G. Adamo, S. Stivala, A. Parisi, P. Livreri, A.C. Busacca, Hydrogel films engineered in a mesoscopically ordered structure and responsive to ethanol vapors, React. Funct. Polym. 79 (2014) 68–76. https://doi.org/10.1016/j.reactfunctpolym.2014.03.016

[14] L. Burratti, F. De Matteis, M. Casalboni, R. Francini, R. Pizzoferrato, P. Prosposito, Polystyrene photonic crystals as optical sensors for volatile organic compounds, Mater. Chem. Phys. 212 (2018) 274–281. https://doi.org/10.1016/j.matchemphys.2018.03.039

[15] L. Burratti, M. Casalboni, F. De Matteis, R. Pizzoferrato, P. Prosposito, Polystyrene opals responsive to methanol vapors, Materials (Basel). 11 (2018). https://doi.org/10.3390/ma11091547

[16] M. Qin, M. Sun, R. Bai, Y. Mao, X. Qian, D. Sikka, Y. Zhao, H.J. Qi, Z. Suo, X. He, Bioinspired Hydrogel Interferometer for Adaptive Coloration and Chemical Sensing, Adv. Mater. 30 (2018) 1800468. https://doi.org/10.1002/adma.201800468

[17] J. Owen, L. Brus, Chemical Synthesis and Luminescence Applications of Colloidal Semiconductor Quantum Dots, J. Am. Chem. Soc. 139 (2017) 10939–10943. https://doi.org/10.1021/jacs.7b05267

[18] K.J. Nordell, E.M. Boatman, G.C. Lisensky, A Safer, Easier, Faster Synthesis for CdSe Quantum Dot Nanocrystals, J. Chem. Educ. 82 (2005) 1697. https://doi.org/10.1021/ed082p1697

[19] F. Liu, M.-H. Jang, H.D. Ha, J.-H. Kim, Y.-H. Cho, T.S. Seo, Facile Synthetic Method for Pristine Graphene Quantum Dots and Graphene Oxide Quantum Dots: Origin of Blue and Green Luminescence, Adv. Mater. 25 (2013) 3657– 3662. https://doi.org/10.1002/adma.201300233

[20] Q. Lu, C. Wu, D. Liu, H. Wang, W. Su, H. Li, Y. Zhang, S. Yao, A facile and simple method for synthesis of graphene oxide quantum dots from black carbon, Green Chem. 19 (2017) 900–904. https://doi.org/10.1039/C6GC03092K

[21] L. Burratti, E. Ciotta, E. Bolli, M. Casalboni, F. De Matteis, R. Francini, S. Casciardi, P. Prosposito., Synthesis of fluorescent silver nanoclusters with potential application for heavy metal ions detection in water, in AIP Conference Proceedings; (2019): p. 020007. https://doi.org/10.1063/1.5123568

[22] L. Burratti, E. Bolli, M. Casalboni, F. de Matteis, F. Mochi, R. Francini, S. Casciardi, P. Prosposito, Synthesis of Fluorescent Ag Nanoclusters for Sensing and Imaging Applications, Mater. Sci. Forum. 941 (2018) 2243–2248. https://doi.org/10.4028/www.scientific.net/MSF.941.2243

[23] A.C. Vinayaka, S. Basheer, M.S. Thakur, Bioconjugation of CdTe quantum dot for the detection of 2,4-dichlorophenoxyacetic acid by competitive fluoroimmunoassay based biosensor, Biosens. Bioelectron. 24 (2009) 1615–1620. https://doi.org/10.1016/j.bios.2008.08.042

[24] R. De Angelis, L. D'Amico, M. Casalboni, F. Hatami, W.T. Masselink, P. Prosposito, Photoluminescence sensitivity to methanol vapours of surface InP quantum dot: Effect of dot size and coverage, Sensors Actuators, B Chem. 189 (2013) 113–117. https://doi.org/10.1016/j.snb.2013.01.057

[25] M. Frasco, N. Chaniotakis, Semiconductor Quantum Dots in Chemical Sensors and Biosensors, Sensors. 9 (2009) 7266–7286. https://doi.org/10.3390/s90907266

[26] A. Ananthanarayanan, X. Wang, P. Routh, B. Sana, S. Lim, D.-H. Kim, K.-H. Lim, J. Li, P. Chen, Facile Synthesis of Graphene Quantum Dots from 3D Graphene and their Application for Fe 3+ Sensing, Adv. Funct. Mater. 24 (2014) 3021–3026. https://doi.org/10.1002/adfm.201303441

[27] E. Ciotta, P. Prosposito, P. Tagliatesta, C. Lorecchio, L. Stella, S. Kaciulis, P. Soltani, E. Placidi, R. Pizzoferrato, Discriminating between different heavy metal ions with fullerene-derived nanoparticles, Sensors (Switzerland). 18 (2018) 1–15. https://doi.org/10.3390/s18051496

[28] J.X. Dong, Z.F. Gao, Y. Zhang, B.L. Li, N.B. Li, H.Q. Luo, A selective and sensitive optical sensor for dissolved ammonia detection via agglomeration of fluorescent Ag nanoclusters and temperature gradient headspace single drop microextraction, Biosens. Bioelectron. 91 (2017) 155–161. https://doi.org/10.1016/j.bios.2016.11.062

[29] A.T. Afaneh, G. Schreckenbach, Fluorescence Enhancement/Quenching Based on Metal Orbital Control: Computational Studies of a 6-Thienyllumazine- Based Mercury Sensor, J. Phys. Chem. A. 119 (2015) 8106–8116. https://doi.org/10.1021/acs.jpca.5b04691

[30] S. O'Keeffe, C. Fitzpatrick, E. Lewis, An optical fibre based ultra violet and visible absorption spectroscopy system for ozone concentration monitoring, Sensors Actuators B Chem. 125 (2007) 372–378. https://doi.org/10.1016/j.snb.2007.02.023

[31] H.-A. Ho, M. Béra-Abérem, M. Leclerc, Optical Sensors Based on Hybrid DNA/Conjugated Polymer Complexes, Chem. - A Eur. J. 11 (2005) 1718– 1724. https://doi.org/10.1002/chem.200400537

[32] J. Wang, Y. Chang, W.B. Wu, P. Zhang, S.Q. Lie, C.Z. Huang, Label-free and selective sensing of uric acid with gold nanoclusters as optical probe, Talanta. 152 (2016) 314–320. https://doi.org/10.1016/j.talanta.2016.01.018

[33] K.A. Willets, R.P. Van Duyne, Localized Surface Plasmon Resonance Spectroscopy and Sensing, Annu. Rev. Phys. Chem. 58 (2007) 267–297. https://doi.org/10.1146/annurev.physchem.58.032806.104607

[34] K.M. Mayer, J.H. Hafner, Localized Surface Plasmon Resonance Sensors, Chem. Rev. 111 (2011) 3828–3857. https://doi.org/10.1021/cr100313v

[35] J. V. Rohit, J.N. Solanki, S.K. Kailasa, Surface modification of silver nanoparticles with dopamine dithiocarbamate for selective colorimetric sensing of mancozeb in environmental samples, Sensors Actuators, B Chem. 200 (2014) 219–226. https://doi.org/10.1016/j.snb.2014.04.043

[36] D. Li, Y. Dong, B. Li, Y. Wu, K. Wang, S. Zhang, Colorimetric sensor array with unmodified noble metal nanoparticles for naked-eye detection of proteins and bacteria, Analyst. 140 (2015) 7672–7677. https://doi.org/10.1039/c5an01267h

[37] A. Jeevika, D.R. Shankaran, Functionalized silver nanoparticles probe for visual colorimetric sensing of mercury, Mater. Res. Bull. 83 (2016) 48–55. https://doi.org/10.1016/j.materresbull.2016.05.029

[38] J.Y. Cheon, W.H. Park, Green synthesis of silver nanoparticles stabilized with mussel-inspired protein and colorimetric sensing of lead(II) and copper(II) ions, Int. J. Mol. Sci. 17 (2016). https://doi.org/10.3390/ijms17122006

[39] P. Prosposito, F. Mochi, E. Ciotta, M. Casalboni, F. De Matteis, I. Venditti, L. Fontana, G. Testa, I. Fratoddi, Hydrophilic silver nanoparticles with tunable optical properties: Application for the detection of heavy metals in water, Beilstein J. Nanotechnol. 7 (2016) 1654–1661. https://doi.org/10.3762/bjnano.7.157

[40] F. Mochi, L. Burratti, I. Fratoddi, I. Venditti, C. Battocchio, L. Carlini, G. Iucci, M. Casalboni, F. De Matteis, S. Casciardi, S. Nappini, I. Pis, P. Prosposito, Plasmonic Sensor Based on Interaction between Silver Nanoparticles and Ni2+ or Co2+ in Water, Nanomaterials. 8 (2018) 488. https://doi.org/10.3390/nano8070488

[41] A. Majzik, L. Fülöp, E. Csapó, F. Bogár, T. Martinek, B. Penke, G. Bíró, I. Dékány, Functionalization of gold nanoparticles with amino acid, β-amyloid peptides and fragment,

Colloids Surfaces B Biointerfaces. 81 (2010) 235–241. https://doi.org/10.1016/j.colsurfb.2010.07.011

[42] I. Venditti, G. Testa, F. Sciubba, L. Carlini, F. Porcaro, C. Meneghini, S. Mobilio, C. Battocchio, I. Fratoddi, Hydrophilic Metal Nanoparticles Functionalized by 2-Diethylaminoethanethiol: A Close Look at the Metal– Ligand Interaction and Interface Chemical Structure, J. Phys. Chem. C. 121 (2017) 8002–8013. https://doi.org/10.1021/acs.jpcc.7b01424

[43] Prosposito, Burratti, Bellingeri, Protano, Faleri, Corsi, Battocchio, Iucci, Tortora, Secchi, Franchi, Venditti, Bifunctionalized Silver Nanoparticles as Hg2+ Plasmonic Sensor in Water: Synthesis, Characterizations, and Ecosafety, Nanomaterials. 9 (2019) 1353. https://doi.org/10.3390/nano9101353

[44] L. Li, L. Gui, W. Li, A colorimetric silver nanoparticle-based assay for Hg(II) using lysine as a particle-linking reagent, Microchim. Acta. 182 (2015) 1977– 1981. https://doi.org/10.1007/s00604-015-1536-2

[45] P.K. Sarkar, A. Halder, N. Polley, S.K. Pal, Development of Highly Selective and Efficient Prototype Sensor for Potential Application in Environmental Mercury Pollution Monitoring, Water, Air, Soil Pollut. 228 (2017) 314. https://doi.org/10.1007/s11270-017-3479-1

[46] G. V Ramesh, T.P. Radhakrishnan, A Universal Sensor for Mercury (Hg, Hg I , Hg II) Based on Silver Nanoparticle-Embedded Polymer Thin Film, ACS Appl. Mater. Interfaces. 3 (2011) 988–994. https://doi.org/10.1021/am200023w

[47] S.S. Ravi, L.R. Christena, N. SaiSubramanian, S.P. Anthony, Green synthesized silver nanoparticles for selective colorimetric sensing of Hg2+ in aqueous solution at wide pH range, Analyst. 138 (2013) 4370. https://doi.org/10.1039/c3an00320e

Two Different Acid Oxidation Syntheses to Open C_{60} Fullerene for Heavy Metal Detection

E. Ciotta[1*], L. Burratti[1], P. Prosposito[2], E. Bolli[3], S. Kaciulis[3], S. Antonaroli[4] and R. Pizzoferrato[1]

[1]Department of Industrial Engineering University of Rome Tor Vergata, 00133 Rome, Italy

[2] Department of Industrial Engineering INSTM and CiMER University of Rome Tor Vergata, 00133 Rome, Italy

[3]Institute for the Study of Nanostructured Materials, CNR of Italy, Monterotondo Stazione, 00015 Rome, Italy

[4]Department of Chemical Science and Technologies, University of Rome Tor Vergata, 00133 Rome, Italy

* erica.ciotta@uniroma2.it

Keywords: Carbon Materials, Heavy Metals, Sensors, Spectroscopy, Photoluminescence, Quenching, Chemical Oxidation

Abstract. Graphene oxide quantum dots (GOQDs) can be synthesized through a large variety of synthesis methods starting from different carbon allotropes such as nanotubes, graphite, C_{60} and exploiting various synthesis and reactions. These different approaches have great influence on the properties of the obtained materials, and, consequently, on the potential applications. In this work, Buckminster C_{60} fullerene has been used to prepare unfolded fullerene nanoparticles (UFNPs) via two distinct synthesis methods namely: Hummer and $H_2SO_4 + HNO_3$ solution. The different characteristics of the final materials and the different response in the presence of heavy metal ions have been investigated in view of sensing applications of water contamination.

Introduction

Graphene oxide quantum dots present unique properties [1] of photoluminescence (PL), biocompatibility, low cytotoxicity and photostability. For all these reasons they have been proposed as candidate material for many applications such as bioimaging and electrochemical biosensor, catalysis, organic light-emitting diodes, and heavy metal detection [2-5]. GOQDs are single or few-layer graphene oxide (GO) sheets with lateral dimensions less than 100 nm [6] exhibiting exciton confinement and quantum-size effect. In a typical structure the presence of oxygen-containing functional groups (–OH, –COOH and epoxy groups) located on the carbon basal plane and at the edges of the sheets are quite common. The consequent large number of sp^3 carbon atoms in the GO lattice opens the optical bandgap of graphene and results in photo-excited fluorescence with a typical, stable blue/green emission which can be exploited for many possible applications [7]. The main role of these groups on the surface, which either form during the synthesis procedures or are intentionally added by specific treatments, consists in improving the hydrophilicity and thus the dispersibility in water. More importantly, they can undergo partial deprotonation in water environment, and this can promote binding of heavy metal cations as substitutes for the lost protons [8,9]. For this reason, GOQDs have been used as sensor of heavy

metals (HMs) ions in water. In fact, chelation of heavy metals has recently been considered as the origin of the fluorescence quenching observed in (GOQDs) [4,10-16]
Although GO is obtained starting from smaller molecules [17] with the so called bottom-up approach, they are usually produced starting from macroscopic materials like graphite, nanotubes, or C_{60} trough a top-down method. The most common synthesis of this type are: acid oxidation [18], electrochemical oxidation [19], hydrothermal [20], microwave radiation [21], chemical exfoliation [22]. In the case of acid oxidation, there are different ways to obtain graphene oxide. In the Staudenmaier method [23], for example, a combination of sulfuric acid and fuming nitric acid and $KClO_3$ are used. In the Hofmann method [24], a combination of sulfuric acid (H_2SO_4), concentrated nitric acid (HNO_3), and potassium chlorate ($KClO_3$) are exploited.

The Hummer method [25], based on a combination of sulfuric acid and $NaNO_3$ and $KMnO_4$ or a mixture between sulfuric acid and nitric acid, represents the most common strategy to obtain GOQDs. This strategy is preferred with respect to other methods due to its higher degree of oxidation [26].

Recently, several groups have applied GOQDs produced in the above mentioned synthesis to the study of the interaction with heavy metal ions.

Different responses were obtained depending on the starting materials, the dimension of GOQDs, the number of functional groups on the edge and on the surface and so on. For example, responses to a number of heavy metal ions have been analyzed starting by exfoliated graphite using modified Hummer method [4,11], or by using pyrolysis of citric acid [12] or chemical oxidation of carbon fibers [16], or Hummer method starting from C_{60} fullerene [27,28]. The results were sometime not comparable and a dependence on synthetic method of production of GOQDs was observed that suggest the need of further studies.

In this work we reported on two different synthesis methods starting from C_{60} fullerene to obtain unfolded fullerene nanoparticles (UFNPs) that are a specific type of GOQDs. The first approach is based on a slight modification of Hummer method, while the second one exploited an acid oxidation synthesis with sulphuric and nitric acid. In both cases, the chemical oxidation of C_{60} caused the opening of the Buckminster structure and the formation of oxygen-containing functional groups, such as –OH and –COOH on the edge and the surface of the structure. Both obtained photoluminescent materials were tested to detect the presence in water of heavy metal ions through the change of its PL intensity. We reported how the optical response depends on the two synthesis methods implemented. Moreover, exploiting X-ray photoemission spectroscopy (XPS) and Fourier Transform Infrared Spectroscopy (FTIR), we analyzed how the observed differences can be related to the morphology and chemical properties of the materials.

Experimental

Materials

C_{60} (Solaris Chem Inc.), sulfuric acid (H_2SO_4, Sigma Aldrich 99.9%), sodium nitrate ($NaNO_3$, Sigma Aldrich), sodium hydroxide (NaOH, Sigma Aldrich), potassium permanganate ($KMnO_4$, Sigma Aldrich), hydrogen peroxide (H_2O_2, Sigma Aldrich), Nitric acid (HNO_3 Sigma Aldrich 99.9%) were used as received without further purification. Deionized water obtained from a Milli-Q water purification system (Millipore) was used for all the synthesis and the sensing test measurements.

<u>*Hummer Synthesis (Synthesis A)*</u>

In the case of Hummer synthesis, hereafter referred to as synthesis A, UFNPs were prepared starting by fullerene C_{60} with a modification of Hummer's synthesis using sodium nitrate, sulphuric acid and potassium permanganate as strong oxidant agents following the recipe of reference [18]. The solution with C_{60}, sodium nitrate and sulphuric acid was stirred in an ice bath for 2 hours, while potassium permanganate was added gradually. Then the solution was further stirred for 4 hours at room temperature. After this time, the temperature was raised to 70°C and then lowered to 35°C and addition of water was made. The reaction was stopped with $H_2O_2$3% and the pH was brought to 7 with NaOH 1 M. The product was dialyzed in a dialysis bag of 2000 Da to remove the residual salts

Synthesis H_2SO_4 + HNO_3 (Synthesis B)

In the case of the synthesis with sulphuric and nitric acid, hereafter referred to as synthesis B, UFNPs were prepared starting by fullerene C_{60} and using concentrated acids H_2SO_4 and HNO_3 in a ratio 3:1. The precursors were mixed together and sonicated for 2 hours. After this time, the solution was put in an oil bath at 150°C for 24 hours. At the end, the solution was cooled down and basified to pH 7 using NaOH 15 M. The product was filtered with 0.22 μm and then dialyzed in a dialysis bag of 2000 Da to remove the residual salts.

Apparatus and data processing

Both solutions were characterized by UV-Vis absorption measurements (ABS) and photoluminescence (PL) by using quartz cuvettes with 1 cm optical path. The absorption spectra were taken with a Cary 50 spectrophotometer in the range 200 – 500 nm. The photoluminescence spectra were taken in the range 350 – 700 nm by exciting the solution with a 200-W continuous Hg(Xe) discharge lamp by a conventional 90-degree geometry. To take the spectra, apposite optical filters were used both for the excitation light and the PL signal. Fourier transform infrared spectroscopy (FT-IR) data were obtained on sample powders by a Perkin Elmer Spectrum One spectrometer (Waltham, MA, USA). In the case of XPS the surface analyses were carried in an electronic spectrometer EscalabMkII (VG Scientific Ltd., East Grinstead, UK) equipped with XPS and AES techniques. The Al Kα source was used for the photoemission and X-ray induced Auger spectroscopy, where as a LEG200 electron gun was used for the AES. The photoemission spectra were collected at 50 eV pass energy in selected area mode A3x10 with a diameter of 3 mm, whereas the Auger spectra were registered at the pass energy of 100 eV, in order to increase the signal-to-noise ratio. All experimental data were processed by using the software Avantage v.5 (Thermo Fisher Scientific, EastGrinstead, UK). Experimental C KVV spectra were smoothed at least for 11 times by moving average routine with a width of 1.2 eV. Afterwards, these spectra were differentiated by using a width of 7 data points for the determination of D parameter

Results

Characterizations of UFNPs

The PL and the absorption spectra of the two systems present some differences as reported in figure 1a. When excited at 300 nm, the PL spectrum of sample obtained from synthesis A has a maximum at 450 nm while sample obtained from synthesis B presents a red shifted peak with maximum at 500 nm and a wider shape with respect to the previous one. The absorption spectra of both solutions are reported in figure 1b. They present the typical peaks of the π-π* transitions around 200 nm and n-π* transition at 300 nm.

Materials Research Forum LLC
https://doi.org/10.21741/9781644900710-3

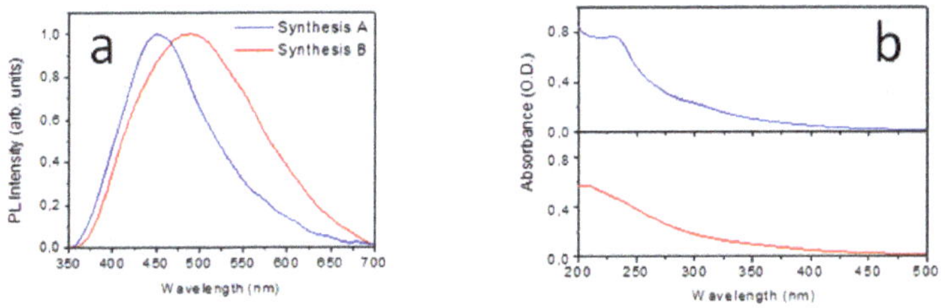

Figure 1 a) PL Intensity spectra of Synthesis A (blue line) and Synthesis B (red line);b) Absorption spectra of Synthesis A (blue line) and Synthesis B (red line)

The π-π* transitions, is due to the presence of the aromatic sp^2 domains (C=C), and the n-π* is generally attributed to the presence of C=O bonds of oxygen-containing functional groups [29]. These groups are typical of graphene oxide quantum dots (GOQDs).

In order to confirm the formation of GOQDs FT-IR spectra were acquired and shown in Figure 2.

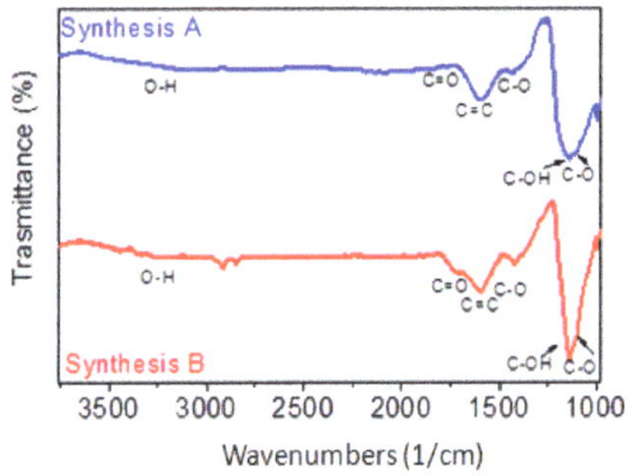

Figure 2 FT-IR spectra of Synthesis A (blue line) and Synthesis B (red line)

In both FT-IR spectra the typical peaks of graphene oxide that are related to the carboxyl, hydroxyl, and epoxy groups of GOQDs are present. In particular, the peaks at approximately 1,000 cm^{-1}, 1,150 cm^{-1}, 1,650 cm^{-1}, 1,750 cm^{-1}, and 3,350 cm^{-1} corresponding to C-O, C-OH, C=C, C=O and O-H, respectively [30].

As a general remark of the characterizations illustrated in Fig1 and Fig 2, we pointed out that, the PL of the two solutions indicates different dimension of the nanoparticles and different number of carboxylic groups. Both these factors are strongly related to the methods of synthesis and to the starting compounds [31].

It is therefore probable that a change in the number of carboxyl groups in the structure leads to a difference response in the presence of heavy metals ions.

A greater number of functional groups allows a greater interaction between GOQDs and ions.

Materials Research Forum LLC
https://doi.org/10.21741/9781644900710-3

The XPS spectra of C 1s region reported in Figs. 3(a) and 3(b) testified the prevalence of C–C bonds (main peak named as A in the Fig 3 (a) and (b)) with binding energy BE = 284.6 eV. Other two minor components of C 1s peak, attributed to C–O bonds with BE = 286.4 eV and carboxylic bonds with BE = 288.6 eV, are higher in the sample of synthesis B (Fig. 3 (b)).

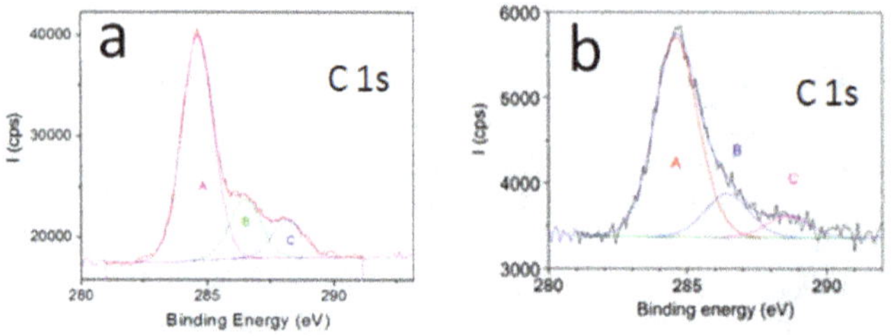

Figure 3 XPS spectra of carbon related to the samples of synthesis A (a) and synthesis B (b).

From the comparison of Auger spectra of C KVV region (Figure 4) acquired by using X-rays and electron beam (see Figure 4 (a)), were determined the values of D parameter equal to 12.7 and 20.9 eV, respectively. This change of the D parameter from the diamond-like value obtained with X-rays to graphitic one (electron beam) indicates that the main configuration of C–C bonds in the sample of synthesis A corresponds to graphene [32]. In the sample of synthesis B, the D parameter obtained from X-rays excited C KVV spectrum (Fig. 4(b)) is higher (15.1 eV) due to the higher amount of C-O and carboxylic bonds.

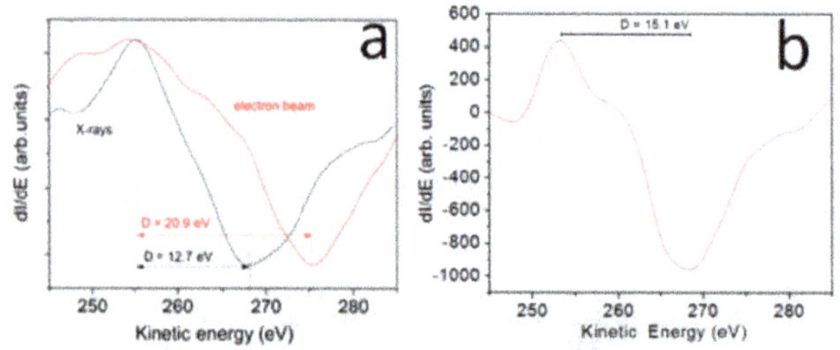

Figure 4 Auger spectra of C KVV region of the samples of synthesis A (a) and synthesis B (b).

<u>Excitation spectra of UFNPs</u>
GOQDs have localized sp2 carbon subdomains having specific absorption energy levels with similar DOS (density of states) as reported by many works [33,34]. These energy states were studied by different authors [33,34] and recently their presence was confirmed by a red-blue-red (RBR) shift of the PL spectra using different excitation wavelengths [35,36].

In figure 5 the emission spectra at various excitation wavelengths of the two systems are reported. For synthesis A (Figure 5(a)), the emission spectra are reported in the excitation range

270 – 360 nm, while, for synthesis B are reported in the range 280 – 400 nm (Figure 5(b)). As it is possible to see from the graph, in the synthesis B case, the maximum of photoluminescence spectrum has an RBR shift [35,36]. Specifically, a redshift is present in the excitation wavelength range 280–290 nm, subsequently a blueshift in 300–340 nm, and a redshift in the range 340–500 nm. In the case of synthesis A, this shift is no present.

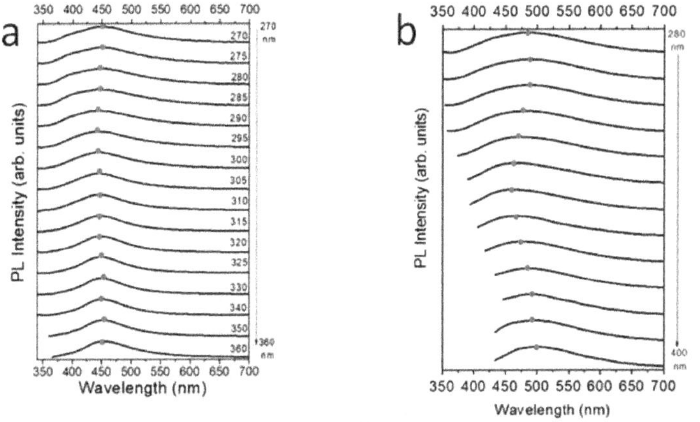

Figure 5 a) Emission spectra of synthesis A in the excitation range 270 – 360 nm; b) Emission spectra of synthesis B in the excitation range 280 – 400 nm

Called intrinsic state the state formed by localized sp2 carbon subdomains, and extrinsic states the state due to the presence of the oxygen-containing groups [35,36], the difference in the emission spectra at various excitation can be related to the number variation of intrinsic and extrinsic states. The RBR shift is present when the intrinsic and extrinsic states are in greater numbers, in fact, for synthesis B the number of oxygen-containing groups is greater than the synthesis A, and this is due to the synthesis method. In fact, a different synthesis method causes a variation of defective groups, thus a variation in the number of intrinsic and extrinsic groups. For this reason, the RBR shift is present for the synthesis B.

Heavy metal sensitivity

We measured the optical response of the both synthesized systems to test the presence of heavy metal ions in water. The ratio of the fluorescence intensity of the two UFNPs syntheses, with and without of metal ions in the solution is reported in figure 6 for several metal ions.

In order to characterize the system response in the presence of these ions at the concentration of 100 μM, a drop of 30 μl of concentrated salt solution was added to 3 ml of UFNPs solution. The pH of the solution was measured immediately before and after each measurement and it was 7. The investigated ions were: Co(II), Cu(II), Cd(II), Pb(II), Ni(II), Na(I), As(III), and As(V).

Materials Research Forum LLC
https://doi.org/10.21741/9781644900710-3

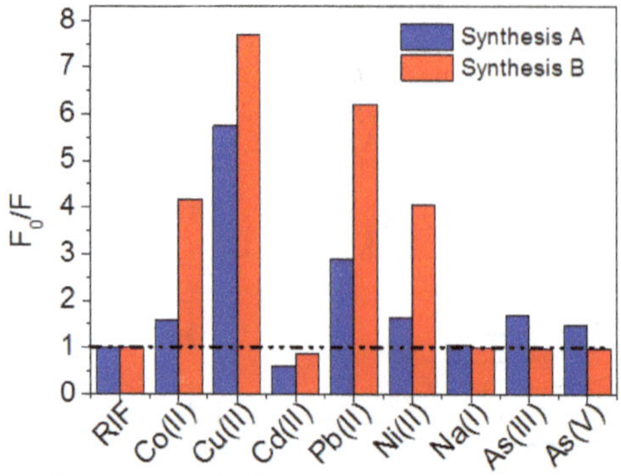

Figure 6 Response in the presence of different heavy metal ions at 100 μM concentration for Synthesis A (blue bar) and Synthesis B (red bar)

As stated in our previous works [27,28,37] we reported different studies about the interaction between UFNPs and heavy metal ions. Due to an aggregation process, the interaction of heavy metal ions and UFNPs can cause the formation of a stable complex [37,38]. In fact, metal ions can bind more quantum dots together thanks to the presence of the oxygen-containing groups at the edge and at the surface of the UFNPs structure [38-40]. This aggregation is responsible for a change of the PL intensity.

The binding with a specific ion depends by the affinity between the ion and the UFNPs. This mechanism plays an important role in the Photoinduced Electron Transfer (PET) in the UFNPs and therefore, the presence of some ions causes a quenching or an enhancement of the PL signal. The UFNPs synthesized with Hummer method shows a fluorescence quenching and an increase of absorption in the short wavelength range of absorbance spectrum (not shown here, see reference [27]) in the presence of Cu(II) and Pb(II) ions, as we reported recently [27]. This has been explained with the interaction of metal ions with the carboxyl groups located at the edges of UFNPs, possibly promoting an edge-to-edge aggregation. In the case of the arsenic ions, the aggregation could be face-to-face [40]. Both types of aggregation allow the chelation of metal ions, with a consequence of PL quenching effect. In the case of cadmium ions, the PL enhanced could be due to the Chelation-Enhanced Fluorescence (CHEF) mechanism [40,41], Cd(II) ions might coordinate with the basal surface of the carbon layers trough cation-π interaction [41-43] using the oxygen lone pair electrons of the carbonyl and epoxy groups by immobilizing the lone pair electrons, Cd(II) cation increases the recombination rate and the fluorescence intensity of UFNPs.

In the figure 5 both syntheses have a similar behaviour in the presence of the different ions, but in the case of synthesis B, the system is more sensitive. This is probably due to a greater oxidation degree of the system, in fact, FT-IR spectra (Figure 2), XPS (Figure 3) and Auger spectroscopy (Figure 4) confirm the greater presence of carboxylic groups. Or probably, it is due the different synthesis method, without potassium permanganate. In fact, the use of potassium permanganate can limit the response of the system, due to possible interaction between Mn(II) ions and quantum dots. Not using the permanganate in the synthesis could allow to obtain a more sensitive system.

An important aspect about the response of the two systems in presence of the arsenic ions is reported: As(III) and As(V) not affect the optical signal in the synthesis B, while the both As ions increase the PL emission of the synthesis A. Such a behaviour is probably related to the oxidative groups that are mainly at the edge of the GOQDs, while the arsenic is known to bind preferentially with the surface of the carbon nanosheets, rather than the edge [40,41]. Thanks to the different response in the case of arsenic ions by using both UFNPs solutions it could be possible to detect selectively these ions in water.

Conclusions

A new synthesis of UFNPs starting from C_{60} fullerene was studied by using only two strong acid as oxidant agents. The obtained material was studied and tested to detect different heavy metal ions in water, and the results were compared with our previous synthesis of UFNPs by a modification of Hummer method. A different oxidation degree of carbon is confirmed by FT-IR analysis and XPS. The different structure obtained have different number of extrinsic and intrinsic states and for this reason there is a different response in the emission spectra at various excitation wavelength. The second system is more sensitive to the presence of heavy metal ions, but very interesting in the different response in the presence of As(III) and As(V). In fact, in the case of the second synthesis, the system doesn't respond to the presence of these ions on the contrary of synthesis A. Further studies are necessary to understand this behavior, but these two systems can be used to detect the presence of arsenic in water.

Acknowledgments

This research was funded by Regione Lazio, through Progetto di ricerca 85-2017-15125, according to L.R.13/08 and by the University of Rome Tor Vergata in the framework of "GHOST" project, within "Mission Sustainability" program (D.R. 2817/2016), grant number (CUP): E86C18000450005.

References

[1] X. T. Zheng, A. Ananthanarayanan, K. Q. Luo, and P. Chen, "Glowing Graphene Quantum Dots and Carbon Dots : Properties , Syntheses , and Biological Applications" Small, vol. 11, no. 14, pp. 1620–1636, 2015. https://doi.org/10.1002/smll.201402648

[2] L. Fan, Y. Hu, X. Wang, L. Zhang, F. Li, D. Han, L. Zhenggang, Z. Qixian, Z. Wang and L. Niu, "Fluorescence resonance energy transfer quenching at the surface of graphene quantum dots for ultrasensitive detection of TNT" Talanta, vol. 101, pp. 192–197, 2012. https://doi.org/10.1016/j.talanta.2012.08.048

[3] M. K. Kumawat, M. Thakur, R. B. Gurung, and R. Srivastava, "Graphene Quantum Dots for Cell Proliferation, Nucleus Imaging, and Photoluminescent Sensing Applications" Sci. Rep., vol. 7, no. 1, pp. 1–16, 2017. https://doi.org/10.1038/s41598-017-16025-w

[4] F. Wang, Z. Gu, W. Lei, W. Wang, X. Xia, and Q. Hao, "Graphene quantum dots as a fluorescent sensing platform for highly efficient detection of copper (II) ions" Sensors Actuators B. Chem., vol. 190, pp. 516–522, 2014. https://doi.org/10.1016/j.snb.2013.09.009

[5] E. Zor, E. Morales-Narváez, A. Zamora-Gálvez, H. Bingol, M. Ersoz, and A. Merkoçi, "Graphene quantum dots-based photoluminescent sensor: A multifunctional composite for pesticide detection" ACS Appl. Mater. Interfaces, vol. 7, pp. 20272–20279, 2015. https://doi.org/10.1021/acsami.5b05838

[6] T. Wu, H. Shen, L. Sun, B. Cheng, B. Liu, and J. Shen, "Nitrogen and boron doped monolayer graphene by chemical vapor deposition using polystyrene, urea and boric acid" New J. Chem., vol. 36, no. 6, pp. 1385–1391, 2012. https://doi.org/10.1039/c2nj40068e

[7] Y. Dong, J. Shao, C. Chen, H. Li, R. Wang, Y. Chi, X. Lin and G. Chen, "Blue luminescent graphene quantum dots and graphene oxide prepared by tuning the carbonization degree of citric acid" Carbon N. Y., vol. 50, pp. 4738–4743, 2012. https://doi.org/10.1016/j.carbon.2012.06.002

[8] I. Shtepliuk, N. M. Caffrey, T. Iakimov, V. Khranovskyy, I. A. Abrikosov, and R. Yakimova, "On the interaction of toxic Heavy Metals (Cd, Hg, Pb) with graphene quantum dots and infinite graphene" Sci. Rep., vol. 7, no. 1, p. 3934, 2017. https://doi.org/10.1038/s41598-017-04339-8

[9] R. Sitko, E. Turek, B. Zawisa, E. Malicka, E. Talik, J. Heimann, A. Gagor, B. Feist and R. Wrzalik, "Adsorption of divalent metal ions from aqueous solutions using graphene oxide" Dalt. Trans., vol. 42, no. 16, p. 5682, 2013. https://doi.org/10.1039/c3dt33097d

[10] Jian Ju and W. Chen, "Graphene Quantum Dots as a Fluorescence Probes for Sensing Metal Ions: Synthesis and Applications" Curr. Org. Chem., vol. 19, pp. 1150–1162, 2015. https://doi.org/10.2174/1385272819666150318222547

[11] D. Wang, L. Wang, X. Dong, Z. Shi, and J. Jin, "Chemically tailoring graphene oxides into fluorescent nanosheets for Fe3+ ion detection" Carbon N. Y., vol. 50, no. 6, pp. 2147–2154, 2012. https://doi.org/10.1016/j.carbon.2012.01.021

[12] H. Chakraborti, S. Sinha, S. Ghosh, and S. Kalyan, "Interfacing water soluble nanomaterials with fluorescence chemosensing : Graphene quantum dot to detect Hg2+ in 100 % aqueous solution" Mater. Lett., vol. 97, pp. 78–80, 2013. https://doi.org/10.1016/j.matlet.2013.01.094

[13] Z. Li, Y. Wang, Y. Ni, and S. Kokot, "A rapid and label-free dual detection of Hg (II) and cysteine with the use of fluorescence switching of graphene quantum dots" Sensors Actuators B. Chem., vol. 207, pp. 490–497, 2015. https://doi.org/10.1016/j.snb.2014.10.071

[14] H. Huang, L. Liao, X. Xu, M. Zou, F. Liu, and N. Li, "The electron-transfer based interaction between transition metal ions and photoluminescent graphene quantum dots (GQDs): A platform for metal ion sensing" Talanta, vol. 117, pp. 152–157, 2013. https://doi.org/10.1016/j.talanta.2013.08.055

[15] S. Huang, H. Qiu, F. Zhu, S. Lu, and Q. Xiao, "Graphene quantum dots as on-off-on fluorescent probes for chromium (VI) and ascorbic acid" Microchim. Acta, vol. 182, pp. 1723–1731, 2015. https://doi.org/10.1007/s00604-015-1508-6

[16] X. Liu, W. Gao, and X. Zhou, "Pristine graphene quantum dots for detection of copper ions" J. Mater. Res., vol. 29, no. 13, pp. 1401–1407, 2014. https://doi.org/10.1557/jmr.2014.145

[17] R. Xie, Z. Wang, W. Zhou, Y. Liu, L. Fan, Y. Li, and X. Li, "Graphene quantum dots as smart probes for biosensing" Anal. Methods, vol. 8, no. 20, pp. 4001–4006, 2016. https://doi.org/10.1039/C6AY00289G

[18] C. K. Chua, Z. Sofer, P. Simek, O. Jankovsky, K. Klimova, S. Bakardjieva and M. Pumera , "Synthesis of strongly fluorescent graphene quantum dots by cage-opening buckminsterfullerene" ACS Nano, vol. 9, no. 3, pp. 2548–2555, 2015. https://doi.org/10.1021/nn505639q

[19] X. Tan, Y. Li, X. Li, S. Zhou, L. Fan, and S. Yang, "Electrochemical synthesis of small-sized red fluorescent graphene quantum dots as a bioimaging platform" Chem. Commun., vol. 51, no. 13, pp. 2544–2546, 2015. https://doi.org/10.1039/C4CC09332A

[20] K. Li, W. Liu, Y. Ni, D. Li, D. Lin, Z. Su and G. Wei, “Technical synthesis and biomedical applications of graphene quantum dots” J. Mater. Chem. B, vol. 5, no. 25, pp. 4811–4826, 2017. https://doi.org/10.1039/C7TB01073G

[21] L. L. Li, J. Ji, R. Fei, C. Z. Wang, Q. Lu, J. R. Zhang, and J. J. Zhu, “A Facile Microwave Avenue to Electrochemiluminescent Two-Color Graphene Quantum Dots” Adv. Funct. Mater., vol. 22, no. 14, pp. 2971–2979, 2012. https://doi.org/10.1002/adfm.201200166

[22] F. Liu, M. H. Jang, H. D. Ha, J. H. Kim, Y. H. Cho, and T. S. Seo, “Facile synthetic method for pristine graphene quantum dots and graphene oxide quantum dots: Origin of blue and green luminescence” Adv. Mater., vol. 25, no. 27, pp. 3657–3662, 2013. https://doi.org/10.1002/adma.201300233

[23] L. Staudenmaier, “Darstellung der Graphitslure” Ger. Chem. Soc., vol. 31, no. 2, pp. 1481–1487, 1898. https://doi.org/10.1002/cber.18980310237

[24] H. P. Boehm, A. Clauss, G. O. Fischer, and U. Hofmann, “Dünnste kohlenstoff-folien” Zeitschrift fur Naturforsch. - Sect. B J. Chem. Sci., vol. 17, no. 3, pp. 150–153, 1962. https://doi.org/10.1515/znb-1962-0302

[25] W. S. Hummers and R. E. Offeman, “Preparation of Graphitic Oxide” J. Am. Chem. Soc., vol. 80, no. 6, p. 1339, 1958. https://doi.org/10.1021/ja01539a017

[26] C. Botas P. Álvarez, P. Blanco, M. Granda, C. Blanco, R. Santamaría, and R. Menéndez, , “Graphene materials with different structures prepared from the same graphite by the Hummers and Brodie methods” Carbon N. Y., vol. 65, pp. 156–164, 2013. https://doi.org/10.1016/j.carbon.2013.08.009

[27] E. Ciotta, P. Prosposito, P. Tagliatesta, C. Lorecchio, L. Stella, S. Kaciulis, P. Soltani, E. Placidi and R. Pizzoferrato, “Discriminating between different heavy metal ions with fullerene-derived nanoparticles” Sensors (Switzerland), vol. 18, no. 5, pp. 1–15, 2018. https://doi.org/10.3390/s18051496

[28] E. Ciotta, S. Paoloni, M. Richetta, P. Prosposito, P. Tagliatesta, C. Lorecchio, I. Venditti, I. Fratoddi, S. Casciardi and R. Pizzoferrato, “Sensitivity to Heavy-Metal Ions of Unfolded Fullerene Quantum Dots” Sensors, vol. 17, p. 2614, 2017. https://doi.org/10.3390/s17112614

[29] A. V Naumov, Optical Properties of Graphene Oxide, in Graphene Oxide: Fundamentals and Applications. Chichester, UK: John Wiley & Sons, Ltd, 2016. https://doi.org/10.1002/9781119069447.ch5

[30] L. Yun and Z. Kyusik, “Graphene oxide-modified ZnO particles: synthesis, characterization , and antibacterial properties” Int. J. Nanomedicine, vol. 10, pp. 79–92, 2015. https://doi.org/10.2147/IJN.S88319

[31] S. J. Bradley, R. Kroon, G. Laufersky, M. Röding, R.V. Goreham, T. Gschneidtner, and T. Nann., “Heterogeneity in the fluorescence of graphene and graphene oxide quantum dots” Microchim. Acta, vol. 184, no. 3, pp. 871–878, 2017. https://doi.org/10.1007/s00604-017-2075-9

[32] S. Kaciulis, A. Mezzi, P. Calvani and D.M. Trucchi, "Electron spectroscopy of the main allotropes of carbon", Surf. Interface Anal., vol. 46, pp. 966-969, 2014. https://doi.org/10.1002/sia.5382

[33] P. Huang, J.J. Shi, M. Zhang, X. H. Jiang, H. X. Zhong, Y.M. Ding and J. Lu, "Anomalous Light Emission and Wide Photoluminescence Spectra in Graphene Quantum Dot: Quantum Confinement from Edge Microstructure" J. Phys. Chem. Lett., vol. 7, no. 15, pp. 2888–2892, 2016. https://doi.org/10.1021/acs.jpclett.6b01309

[34] Y. Li, H. Shu, S. Wang, and J. Wang, "Electronic and optical properties of graphene quantum dots: The role of many-body effects" J. Phys. Chem. C, vol. 119, no. 9, pp. 4983–4989, 2015. https://doi.org/10.1021/jp506969r

[35] M. H. Jang, S. H. Song, H. D. Ha, T. S. Seo, S. Jeon, and Y. H. Cho, "Origin of extraordinary luminescence shift in graphene quantum dots with varying excitation energyAn experimental evidence of localized sp2 carbon subdomain" Carbon N. Y., vol. 118, pp. 524–530, 2017. https://doi.org/10.1016/j.carbon.2017.03.060

[36] C. T. Chien, S. Li, W.J. Lai, Y. Yeh, H. A. Chen, I. S. Chen, and M. Chen, "Tunable photoluminescence from graphene oxide" Angew. Chemie - Int. Ed., vol. 51, no. 27, pp. 6662–6666, 2012. https://doi.org/10.1002/anie.201200474

[37] E. Ciotta, P. Prosposito, and R. Pizzoferrato, "Positive curvature in Stern-Volmer plot described by a generalized model for static quenching" J. Lumin., vol. 206, pp. 518–522, 2019. https://doi.org/10.1016/j.jlumin.2018.10.106

[38] H. Huang, L. Liao, X. Xu, M. Zou, F. Liu, and N. Li, "The electron-transfer based interaction between transition metal ions and photoluminescent graphene quantum dots (GQDs): A platform for metal ion sensing" Talanta, vol. 117, pp. 152–157, 2013. https://doi.org/10.1016/j.talanta.2013.08.055

[39] L. Wu, L. Liu, B. Gao, R. Muñoz-Carpena, M. Zhang, H. Chen, and H. Wang,, "Aggregation kinetics of graphene oxides in aqueous solutions: Experiments, mechanisms, and modeling" Langmuir, vol. 29, no. 49, pp. 15174–15181, 2013. https://doi.org/10.1021/la404134x

[40] S. T. Yang , Y. Chang, H. Wang, G. Liu, S. Chen, Y. Wang, and A. Cao, A., "Folding/aggregation of graphene oxide and its application in Cu2+ removal" J. Colloid Interface Sci., vol. 351, no. 1, pp. 122–127, 2010. https://doi.org/10.1016/j.jcis.2010.07.042

[41] X. Yang, L. Xia, and S. Song, "Arsenic Adsorption From Water Using Graphene-Based Materials As Adsorbents: a Critical Review" Surf. Rev. Lett., vol. 24, no. 01, p. 1730001, 2017. https://doi.org/10.1142/S0218625X17300015

[42] A. T. Afaneh and G. Schreckenbach, "Fluorescence Enhancement/Quenching Based on Metal Orbital Control: Computational Studies of a 6-Thienyllumazine-Based Mercury Sensor" J. Phys. Chem. A, vol. 119, no. 29, pp. 8106–8116, 2015. https://doi.org/10.1021/acs.jpca.5b04691

[43] I. Shtepliuk, V. Khranovskyy, and R. Yakimova, "Insights into the origin of the excited transitions in graphene quantum dots interacting with heavy metals in different media" Phys. Chem. Chem. Phys., vol. 19, no. 45, pp. 30445–30463, 2017. https://doi.org/10.1039/C7CP04711H

Materials Research Proceedings **16** (2020) 27-37
https://doi.org/10.21741/9781644900710-4

Nonlinear Optical Characterization of $CsPbBr_3$ Nanocrystals as a Novel Material for the Integration into Electro-Optic Modulators

Francesco Vitale [1,2 a *], Fabio De Matteis [1 b], Mauro Casalboni [1 c], Paolo Prosposito [1 d], Patrick Steglich[2,3 e], Viachaslau Ksianzou [2 f], Christian Breiler [2 g], Sigurd Schrader [2 h], Barbara Paci [4 i] and Amanda Generosi [4 j]

[1] Department of Industrial Engineering, University of Rome Tor Vergata, Via del Politecnico 1, Rome 00133, Italy

[2] Faculty of Engineering and Natural Sciences, Technical University of Applied Sciences Wildau, Wildau D-15745, Germany

[3] IHP-Leibniz Institute for Innovative Microelectronics, Frankfurt (Oder) D-15236, Germany

[4] ISM-CNR-Area di Ricerca di Tor Vergata, Via del Fosso del Cavaliere 100, Rome 00133, Italy

[a] francesco.vitale.mim@gmail.com, [b] fabio.dematteis@uniroma2.it, [c] casalboni@uniroma2.it, [d] paolo.prosposito@uniroma2.it, [e] steglich@ihp-microelectronics.com, [f] viachaslau.ksianzou@th-wildau.de, [g] christian.breiler@th-wildau.de, [h] schrader@th-wildau.de, [i] Barbara.Paci@ism.cnr.it, [j] amanda.generosi@ism.cnr.it

Keywords: $CsPbBr_3$ Nanocrystals, NLO Materials, Z-Scan, Kerr Effect, EOMs

Abstract. The present work is concerned with the investigation of the nonlinear optical response of green emissive $CsPbBr_3$ nanocrystals, in the form of colloidal dispersions in toluene, synthesized via a room-temperature ligand-assisted supersaturation recrystallization (LASR) method. After carrying out a preliminary characterization via X-Ray Diffraction (XRD) and Absorption and Photoluminescence (PL) Spectroscopies, the optical nonlinearity of the as-obtained colloids is probed by means of a single-beam Z-scan setup. Results show that the material in question, within the sensitivity of the experimental apparatus, exhibits a nonlinear refractive index n_2 that is the order of 10^{-15} cm^2/W. Moreover, a three-photon absorption mechanism (3PA) is postulated, according to the fitting of the recorded Z-scan traces and the fundamental absorption threshold, which turns out to be off resonance with twice the energy of the laser radiation. A figure of merit is, then, calculated as an indicator of the quality of the $CsPbBr_3$ nanocrystals as a candidate material for photonic devices, for instance, Kerr-like electro-optic modulators (EOMs).

Introduction

The past decade has been characterized by the flourishing development of fiber-integrated circuits and photonic devices [1] based on silicon-organic-hybrid (SOH) slot waveguides [2], namely two silicon rails separated by a submicrometric-wide slot infiltrated by a nonlinear optical (NLO) medium. Polymers and polymer-dye host-guest systems have been largely employed as active materials suitable for exploiting both the linear and quadratic electro-optic (EO) effects [3] [4] [5]. However it is well-known, that organic systems are usually affected by a remarkable two-photon absorption (2PA) that can diminish the effectiveness of the third-order nonlinearity, similarly to the case of silicon [6]. Moreover, the chemical and thermo-mechanical stability of organic materials is often undermined by the conditions at which the device is required to operate, especially at high temperatures. Hence, the quest for novel and promising nonlinear optical active media that can meet these requirements is ongoing and lead halide perovskite nanocrystals are part of this framework.

Lead halide perovskites have emerged recently as promising materials for many applications in photovoltaics [7] [8] and optoelectronics [9] [10]. As concerns the field of photonics, latest works have shown some novel opportunities for the integration into nonlinear optical devices [11] [12], thanks to the low-cost fabrication and agile processability of these materials. Moreover, they exhibit remarkable electronic and optical properties such as relatively high values of the refractive index, broadband bandgap tunability, large optical gain and strong nonlinear response, which are enhanced at the nanoscale [13].

Experimental Section

Materials. Chemicals are employed as received, according to their original technical grade: CsBr (99.999%, Sigma-Aldrich), PbBr2 (99.999%, Sigma-Aldrich), DMF ($\geq$ 99.5%, Carl Roth), DMSO ($\geq$ 99.5%, Carl Roth), toluene ($\geq$ 99.5%, Carl Roth), oleylamine ($\geq$ 96%, Fisher Scientific) and oleic acid (90%, Fisher Scientific).

Synthesis of $CsPbBr_3$ nanocrystals. $CsPbBr_3$ nanocrystals are synthesized via room-temperature ligand-assisted supersaturation recrystallization (LASR) [14]. A mixed precursor solution is prepared by dissolving 0.04 mmol CsBr in 0.5 mL DMF and 0.04 mmol $PbBr_2$ in 0.5 mL DMSO. Alternatively, the dissolution of salts into the aprotic polar solvents is reversed, namely CsBr in DMSO and $PbBr_2$ in DMF (preparation B). Afterwards, the surface ligands, namely oleylamine (20 μL) and oleic acid (10 μL), are added to the as-obtained 1 mL precursor solution: they help to control the size of the nanocrystals and to disperse them into the anti-solvent. Stirring at 70-75°C for ca. 30 min is accomplished for promoting the complete dissolution of the salts. Hence, 0.25 mL of precursor solution are swiftly injected toluene which acts as the anti-solvent, i.e. a poor solvent for the ions from the precursor salts. The drop between the precursor solubility into the aprotic polar solvents and that into toluene, thus, promotes the formation of a supersaturated state in which the recrystallization of the perovskite phase takes place. Agitation is kept for approximately 1 min, that is the estimated time for the complete formation of nanocrystal [15]. If the vial is irradiated under UV light, a green bright emission is observed, as an indicator for the occurred recrystallization of the perovskite phase.

X-Ray Diffraction Measurements. XRD measurements are performed in reflection mode on a Panalytical Empyrean Diffractometer, using the K_α fluorescence line of a Cu-anode emitting tube. Bragg- Brentano configuration is used as the incident optical pathway (0.25°-0.5°) divergent slits and a solid state hybrid Pix'cel 3D detector, working in 1D linear mode, accomplishes the detection. Samples are studied in the range $10° < 2\theta < 50°$, in the form of thin films drop-cast on glass substrates, with an estimated thickness of few microns.

Spectral Measurements. Absorption spectra are recorded by means of a double beam spectrophotometer (Perkin Elmer Lambda-19), while photoluminescence (PL) measurements are performed via excitation of the 458 nm line of an Ar^+ laser and collected by a compact spectrophotometer (Flame, OceanOptics). Samples are probed in the form of the as-obtained colloids collected inside 1 mm wide glass cuvettes (Hellma® Analytics).

Materials Research Proceedings **16** (2020) 27-37 https://doi.org/10.21741/9781644900710-4

Z-Scan Measurements. The single-beam Z-scan setup is shown in Figure 1. The laser radiation employed for the analysis is the fundamental wavelength (λ = 1064 nm) of a Nd:YAG solid state laser, passively Q-switched via saturable absorber: the nominal pulse duration is $\tau_p \approx$ 30 ps at a repetition rate of $\nu \approx$ 10 Hz. The wavelength chosen for the analysis is assessed to be suitable on account of the fact that the fundamental harmonic of the Nd:YAG is rather adjacent to the telecom wavelengths λ = 1300 nm and λ = 1550 nm. It seems reasonable to expect that the variation of the NLO coefficients in this wavelength range be small, if compared to that for shorter wavelengths.

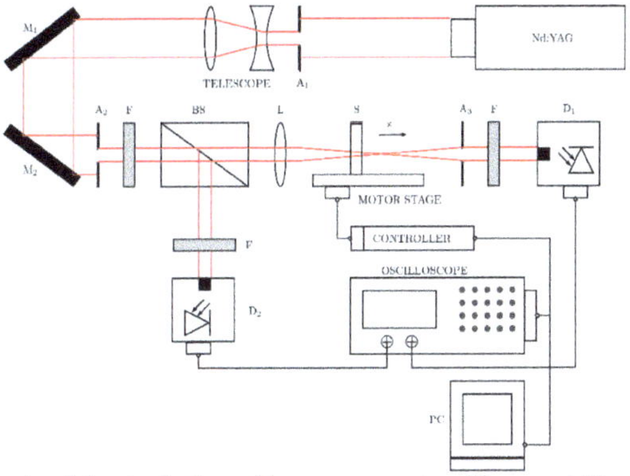

***Figure 1**: Schematic of the single-beam Z-scan apparatus. The acquisition is accomplished through an oscilloscope, whose readout is monitored via LabVIEW. A linear controller interfaced with the PC remotely controls the motor stage: this displaces the sample towards the detector D_1, namely along the positive z-direction.*

The output laser radiation passes through a telescope, which acts as a beam expander and is used for improving the beam quality. Consequently, two mirrors (M_1 and M_2) deflect the beam onto the Z-scan line, where the beam alignment is controlled by the apertures A_2 and A_3. The beam splitter (BS) splits the beam in two parts: ≈ 60% is sent to the probe detector D_1, while ≈ 40% to the reference detector D_2. A converging lens (L) is used for focusing the beam on the sample, while filters (F) are used for attenuating the laser energy and for preventing the saturation of photodetectors, namely, two Si PIN photodiodes with a rise time $t_R \approx$ 35 ns.

Results and Discussion

Spectral and XRD Characterization. A signature for the formation of green emissive $CsPbBr_3$ nanocrystals is represented by the emission peak appearing at 520 nm, as reported in Fig. 2. The fluorescence was excited at 458 nm (edge-filter at 480 nm). The absorption threshold occurring **around 500 nm is usually observed in nanocrystals smaller than 50 nm [15] [16] [17].**

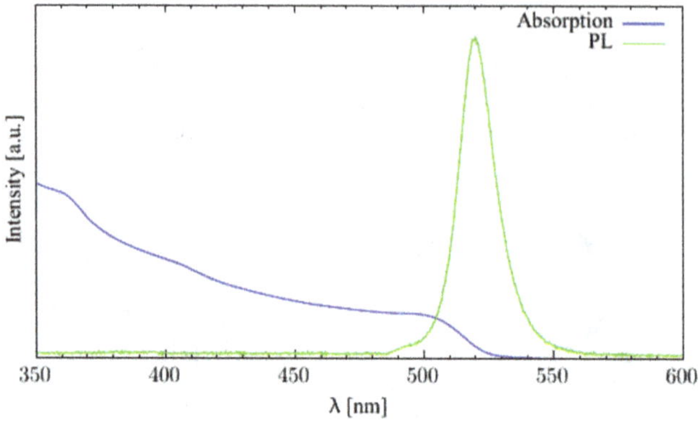

Figure 2*: Typical absorption and PL spectra of green emissive CsPbBr3 NCs colloids prepared via LASR, showing the characteristic absorption threshold occurring at wavelengths around 500 nm, while a quite strong emission is seen to be peaked at 519 nm. Spectra are reported on the same intensity scale only for qualitative purposes.*

The XRD spectra of the perovskite thin films are shown in Fig. 3: both samples present the dominant $CsPbBr_3$ monoclinic phase (ICDD N0. 00- 018- 0364), characterized by lattice constants a = b = 5.82 Å and c = 5.87 Å, and angles $\alpha = \beta = 90°$ and $\gamma = 89.65°$. This phase is characterized by the signature peaks at $2\theta = 15.1°$ and $2\theta = 15.2°$, referred to the diffraction from crystallographic planes (001) and (100) respectively; $2\theta = 21.5°$ and $2\theta = 21.7°$ from planes (110) and (-110); $2\theta = 30.4°$ and $2\theta = 30.7°$, from planes (002) and (200).

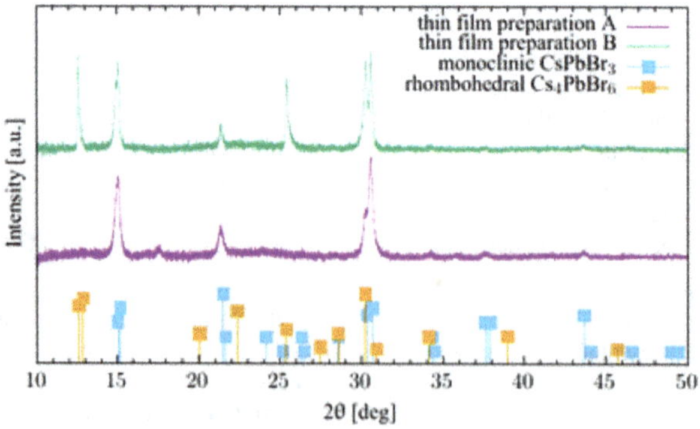

Figure 3*: XRD spectra of thin films from preparations A and B, along with the labeling (colored squares) of XRD reflections from $CsPbBr_3$ monoclinic and rhombohedral Cs_4PbBr_6 phases*

Interestingly, the spectrum referring to preparation B exhibits also fingerprinting peaks of the rhombohedral Cs_4PbBr_6 phase (ICDD No. 01- 073- 2478) with lattice constants a = b = 13.73 Å and c = 17.32 Å, and angles $\alpha = \beta = 90°$ and $\gamma = 120°$, even if the monoclinic phase is still predominant. We attribute this to the higher reactive amount of Cs^+ in preparation B, which tends to react with $PbBr_2$ dissolved in DMF to form Cs_4PbBr_6, confirming the results of Yang et al. [16]. However, only the pure $CsPbBr_3$ NCs colloids obtained from preparation A are devoted to the Z-scan measurements, since, at this point of the analysis, understanding the contribution of the Cs_4PbBr_6 phase to the NLO activity of the nanocrystals is beyond the scope of this work. The average size of the ordered polycrystalline domains has been estimated by means of the Scherrer formula. Peak analysis yields a mean grain size of $D_A = 23 \pm 1$ nm and $D_B = 35 \pm 1$ nm for the $CsPbBr_3$ monoclinic phase resulting from the two preparations, comparable to those reported in [16] [17].

NLO Characterization via Z-Scan: Closed-Aperture Configuration. CS_2 has been used to calibrate the as-built single-beam Z-Scan setup – as originally developed by Sheik-Bahae et al. [18] - in closed-aperture configuration (CA). CS_2 is a standard material for this type of measurements whose nonlinear refractive index is well-known in the literature [19] [20]: $n_2 = (3.1 \pm 0.2) \cdot 10^{-14}$ cm^2/W. To the best of the beam quality optimization, laser energy fluctuations are estimated to be in between 10 - 15%: a cuvette filled with CS_2 is, thus, probed for calibrating the irradiance I_0. Data acquisition is performed by using a 10 cm focal length lens, a diaphragm with aperture radius $r_a \approx 0.5$ mm and linear transmittance $S \approx 0.4$ %, which is sufficiently small to consider the limiting case $S \to 0$ for the CA peak-valley transmittance ΔT_{CA}:

$$\Delta T_{CA} \cong 0.406|\Delta\Phi_0|. \tag{1}$$

$\Delta\Phi_0$ is the time-averaged nonlinear phase shift:

$$\Delta\Phi_0 = k\frac{n_2}{\sqrt{2}}I_0L_{eff} \tag{2}$$

where k is the radiation wavenumber and L_{eff} the optical path inside the cuvette. A z-resolution of $\delta z = 0.5$ mm is found to be a good compromise in between the reproducibility of measurements and the irradiation time that samples are subject to, in order to prevent the potential build-up of thermal effects.

The experimental data are fitted with the theoretical curve for the closed-aperture configuration transmittance [16]:

$$T_{CA}(z,\Delta T) = a + 4\left(\frac{z+b}{z_R}\right)\frac{\Delta T}{0.406}\frac{1}{\left[\left(\frac{z+b}{z_R}\right)^2+9\right]\left[\left(\frac{z+b}{z_R}\right)^2+1\right]} \tag{3}$$

where $\Delta\Phi_0$ is expressed by Eq. (2), z_R is the beam Rayleigh length while a and b are scaling factors: the former accounts for the normalized transmittance offset and the latter for the focus position. Average pulse energy and beam diameter are monitored by using a beam profiler (WinCamD): the beam waist is estimated to be $w_0 = 24 \pm 1$ μm at an irradiance of $I_0 = 14 \pm 2$ GW/cm^2.

Z-scan measurements on solutions and colloids are affected by the solvent effect: thus, it is necessary to accomplish a scan on the solvent as well, in order to subtract any nonlinear contribution from that of the entire solution. CA scans of the solvent (toluene) and the $CsPbBr_3$

Materials Research Proceedings **16** (2020) 27-37 https://doi.org/10.21741/9781644900710-4

NCs colloidal dispersions at three different concentrations (namely 1.5, 2.5 and 4 mM) are shown in Fig. 4.

It must be noticed that when increasing the concentration, scattering processes become predominant: this is apparent from the "humps" emerging in the linear region and, especially, in the asymmetry - with respect to the focus position - affecting the whole Z-scan trace. Fitted ΔT and calculated n_2 values for the perovskite NCs are reported in Tab. 1.

The magnitude and the sign of the NLO coefficients usually depend on several factors:

- Radiation-related: peak power and repetition rate, when too high, usually lead to negative nonlinearities and saturation that can be ascribed to the build-up of thermo-optic effects. Another contribution arises, obviously, from the operating wavelength due to the dispersion of the NLO coefficients.
- Medium-related: degree of crystallinity, morphology and size.

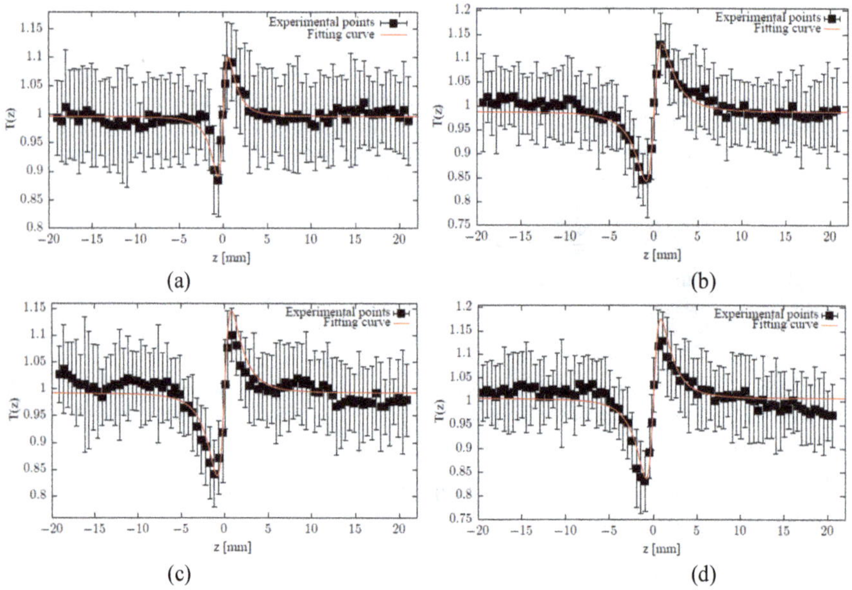

***Figure 4**: CA Z-scan traces of $CsPbBr_3$ NCs colloidal dispersions in toluene at a) 0 mM, b) 1.5 mM, c) 2.5 mM, d) 4 mM.*

As a matter of fact, it is usually difficult to compare NLO coefficients obtained from a Z-scan when not taking into account all of the above-mentioned parameters. To the best of our knowledge, although the literature concerning Z-scan measurements of colloidal $CsPbBr_3$ NCs is still relatively scarce, the n_2 values reported in this work are in good agreement with the results of Liu et al. [21], who found a nonlinear refractive index $n_2 \approx 0.7 \cdot 10\text{-}14$ cm^2/W for green emissive $CsPbBr_3$ nanocrystals (D = 21.4 nm) at 787 nm.

NLO Characterization via Z-Scan: Open-Aperture Configuration In order to investigate NLA, the aperture A_3 placed in front of the detector D_1 is fully opened (S = 100%). Fig. 5 shows the OA Z-scan traces for the perovskite colloidal dispersions at the same three different concentrations as for CA and for toluene.

***Table 1**: ΔT and calculated n_2 values for the perovskite NCs.*

Conc. [mM]	ΔT	$n_{2,P}[\cdot 10^{-14}\ cm^2/W]$
1.5	0.29 ± 0.01	0.4 ± 0.1
2.5	0.31 ± 0.02	0.5 ± 0.1
4	0.34 ± 0.02	0.6 ± 0.1

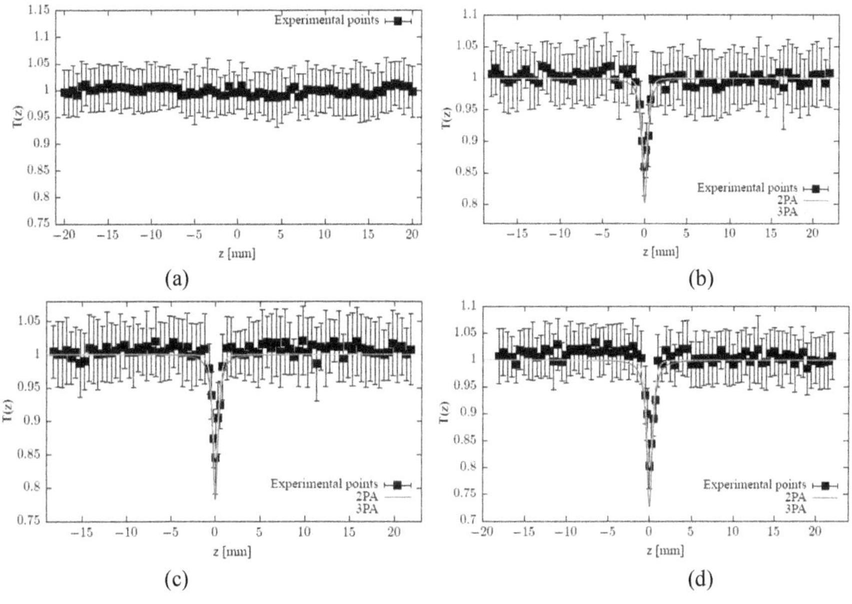

***Figure 5**: OA Z-scan traces of CsPbBr3 NCs colloidal dispersions in toluene at a) 0 mM, b) 1.5 mM, c) 2.5 mM, d) 4 mM.*

Since the solvent exhibits a negligible nonlinear absorption within the sensitivity of the as-built setup, the NLA coefficients returned by the fit, are specifically referred to the perovskite NCs. Experimental data are fitted with the theoretical curves referred to two- and three-photon absorption transmittance [22]:

$$T_{OA}(z, \beta_2) = \sum_{m=0}^{\infty} \frac{(-1)^m}{(m+1)^{\frac{3}{2}}} \left[\frac{\beta_2 I_0 L_{eff}}{1+z^2/z_R^2}\right]^m \quad ; \qquad (4.a)$$

$$T_{OA}(z,\beta_3) = \sum_{m=1}^{\infty} \frac{(-1)^{m-1}}{(2m-1)!(2m-1)^{\frac{1}{2}}} \left[\frac{\sqrt{2\beta_3 I_0^2 L_{eff}^{(2)}}}{1+\frac{(z+b)^2}{z_R}} \right]^{2(m-1)} \quad (4.b)$$

where the effective lengths can be both approximated with the cuvette width at λ = 1064 nm, and β_2 and β_3 represent the 2PA and 3PA coefficients respectively. A three-photon absorption mechanism occurring in perovskite NCs is, thus, postulated, as evidenced from the fact that, within the accuracy guaranteed by the z resolution, the 3PA theoretical curve fits the experimental points better than the 2PA curve. This is reasonable when also considering that the wavelength corresponding to twice the laser energy $\lambda_{2\omega}$ = 532 nm falls outside of the absorption band, while the wavelength corresponding to thrice the laser energy $\lambda_{3\omega} \approx$ 355 nm is clearly inside the resonance region.

However, it must be clarified that in order to corroborate this hypothesis, further investigation is required. For instance, a useful tool is represented by a wavelength-dependent Z-scan analysis, in order to probe the dispersion of the multiphoton absorption coefficients, which follow specific scaling rules as a function of the radiation energy [23] [24]. The β_3 values returned by the fit are shown in Table 2.

***Table 1**: β_3 values for perovskite colloidal dispersions at different concentrations.*

Conc. [mM]	$\beta_{3,P}$[$\cdot 10^{-2}$ cm^3/GW^2]
1.5	1.2 ± 0.2
2.5	1.4 ± 0.2
4	1.8 ± 0.2

A further proof of an apparent 3PA process is given by the work of Manzi et al. [25], who found a 2PA-to-3PA transition at about 1030 nm by studying the nonlinear absorption-induced PL in green emissive $CsPbBr_3$ nanocrystals (D = 10 – 15 nm) by means of a 15ps-pulsed excitation ensured by a tunable laser working in the range 680 – 1080 nm.

Conclusions

We have presented an assessment of the quality of green emissive $CsPbBr_3$ nanocrystals as a potential active medium for electro-optic modulators and other photonic devices exploiting the Kerr effect, on the basis of the NLO coefficients yielded by the Z-scan analysis. Besides the remarkable magnitude of the nonlinear refraction, the postulated 3PA mechanism in the mid-IR range is, indeed, preferable – from a probabilistic point of view - to a 2PA process, which is usually observed in organic materials, within the computation of the optical losses inside a waveguide. Furthermore, the dispersion into an optically-inactive polymer matrix can be also considered as a good strategy for both preventing aggregation mechanism among the perovskite nanocrystals and at the same time improving their stability once deposited into the waveguide.

Acknowledgments

Research supported by Regione Lazio through Progetto di ricerca 85-2017-15125, funded according to L.R.13/08.

References

[1] G. Alimonti, G. Ammendola, A. Andreazza, D. Badoni, V. Bonaiuto, M. Casalboni, F. De Matteis, A. Mai, G. Paoluzzi, P. Prosposito, A. Salamon, G. Salina, E. Santovetti, F. Sargeni, F. Satta, S. Schrader and P. Steglich, "Use of Silicon Photonics Wavelength Multiplexing Techniques for Fast Parallel Readout in High Energy Physics," *Nuclear Instruments and Methods in Physics Research Section A: Accelerators, Spectrometers, Detectors and Associated Equipment,* vol. 936, no. 21, pp. 601-603, 2019. https://doi.org/10.1016/j.nima.2018.09.088

[2] V. Almeida, Q. Xu, C. Barrios and M. Lipson, "Guiding and Confining Light in Void Nanostructure," *Optics Letters,* vol. 29, no. 11, pp. 1209-1211, 2004. https://doi.org/10.1364/OL.29.001209

[3] P. Steglich, C. Mai, D. Stolarek, S. Lischke, S. Kupijai, C. Villringer, S. Pulwer, F. Heinrich, J. Bauer, S. Meister, D. Knoll and M. Casalboni, "Novel Ring Resonator Combining Strong Field Confinement With High Optical Quality Factor," *IEEE Photonics technology Letters,* vol. 27, no. 20, pp. 2197-2200, 2015. https://doi.org/10.1109/LPT.2015.2456133

[4] P. Steglich, "Silicon-On-Insulator Slot Waveguides: Theory and Applications in Electro-Optics and Optical Sensing," in *Emerging Waveguide Technology*, IntechOpen, 2018, pp. 187-210. https://doi.org/10.5772/intechopen.75539

[5] P. Steglich, C. Mai, C. Villringer, S. Pulwer, M. Casalboni, S. Schrader and A. Mai, "Quadratic Electro-Optic Effect in Silicon-Organic Hybrid Slot-Waveguides," *Optics Letters,* vol. 43, no. 15, pp. 3598-3601, 2018. https://doi.org/10.1364/OL.43.003598

[6] J. Leuthold, C. Koos, W. Freude, L. Alloatti, R. Palmer, D. Korn, M. Jazbinsek, J. Pfeifle, M. Lauermann, R. Dinu, M. Waldow, T. Wahlbrink, J. Bolten, M. Fournier, H. Yu, S. Wehrli, J. Fedeli, P. Gunter and W. Bogaerts, "Silicon-Organic Hybrid Electro-Optical Devices," *IEEE Journal of Selected Topics in Quantum Electronics,* vol. 19, no. 6, pp. 114-120, 2013. https://doi.org/10.1109/JSTQE.2013.2271846

[7] H. Snaith, "Perovskites: the Emergence of a New Era for Low-Cost, High-Efficiency Solar Cells," *Journal of Physical Chemistry Letters,* vol. 4, no. 21, pp. 3623-3630, 2013. https://doi.org/10.1021/jz4020162

[8] M. Green, A. Ho-Baillie and H. Snaith, "The Emergence of Perovskite Solar Cells," *Nature Photonics,* vol. 8, no. 7, p. 506, 2014. https://doi.org/10.1038/nphoton.2014.134

[9] F. Deschler , M. Price, S. Pathak, L. Klintberg, D. Jarausch, R. Higler, S. Hüttner, T. Leijtens, S. Stranks, H. Snaith and M. Atatüre, "High Photoluminescence Efficiency and Optically Pumped Lasing in Solution-Processed Mixed Halide," *Journal of Physical Chemistry Letters,* pp. 1421-1426, 2014. https://doi.org/10.1021/jz5005285

[10] S. Stranks and H. Snaith, "Metal-Halide Perovskites for Photovoltaic and Light-Emitting Devices," *Nature Nanotechnology,* vol. 10, no. 5, p. 391, 2015. https://doi.org/10.1038/nnano.2015.90

[11] W. Mao, J. Zheng, C. Zhang, A. Chesman, Q. Ou, J. Hicks, F. Li, Z. Wang, B. Graystone, T. Bell and M. Rothmann, "Controlled Growth of Monocrystalline Organo-Lead Halide

Perovskite and its Application in Photonic Devices," *Angewandte Chemie,* vol. 56, no. 41, pp. 12486-12491, 2017. https://doi.org/10.1002/anie.201703786

[12] A. Chanana, X. Liu, C. Zhang, Z. Vardeny and A. Nahata, "Ultrafast Frequency-Agile Terahertz Devices using Methylammonium Lead Halide Perovskites," Sci. Adv. 4(5), 7353 (2018).," *Science Advances,* vol. 4, no. 5, p. 7353, 2018. https://doi.org/10.1126/sciadv.aar7353

[13] Q. Akkerman, G. Rainò, M. Kovalenko and L. Manna, "Genesis, Challenges and Opportunities for Colloidal Lead Halide Perovskite Nanocrystals," *Nature Materials,* vol. 17, no. 5, p. 394, 2018. https://doi.org/10.1038/s41563-018-0018-4

[14] X. Li, Y. Wu, S. Zhang, B. Cai, Y. Gu, J. Song and H. Zeng, "CsPbX3 Quantum Dots for Lighting and Displays: Room-Temperature Synthesis, Photoluminescence Superiorities, Underlying Origins and White Light-Emitting Diodes," *Advanced Functional Materials,* vol. 26, no. 15, pp. 2435-2445, 2016. https://doi.org/10.1002/adfm.201600109

[15] S. Seth and A. Samanta, "A Facile Methodology for Engineering the Morphology of CsPbX3 Perovskite Nanocrystals under Ambient Condition," *Scientific Reports,* vol. 6, p. 37693, 2016. https://doi.org/10.1038/srep37693

[16] L. Yang, D. Li, C. Wang, W. Yao, H. Wang and K. Huang, "Room-Temperature Synthesis of Pure Perovskite-Related Cs4PbBr6 Nanocrystals and their Ligand-Mediated Evolution into Highly Luminescent CsPbBr3 Nanosheets," *Journal of Nanoparticle Research,* vol. 19, no. 7, p. 258, 2017. https://doi.org/10.1007/s11051-017-3959-7

[17] F. De Matteis, F. Vitale, S. Privitera, E. Ciotta, R. Pizzoferrato, A. Generosi, B. Paci, L. Di Mario, J. Pelli Cresi, F. Martelli and P. Prosposito, "Optical Characterization of Cesium Lead Bromide Perovskites," *Crystals,* vol. 9, p. 280, 2019. https://doi.org/10.3390/cryst9060280

[18] M. Sheik-Bahae, A. Said and E. Van Stryland, "High-Sensitivity, Single-Beam n2 Measurements," *Optics Letters,* vol. 14, no. 17, pp. 955-957, 1989. https://doi.org/10.1364/OL.14.000955

[19] M. Sheik-Bahae, A. Said, H. Wei, D. Hagan and E. Van Stryland, "Sensitive Measurement of Optical Nonlinearities using a Single Beam," *IEEE Journal of Quantum Electronics,* vol. 26, no. 4, pp. 760-769, 1990. https://doi.org/10.1109/3.53394

[20] E. Van Stryland and M. Sheik-Bahae, " Z-Scan Measurements of Optical Nonlinearities," in *Characterization Techniques and Tabulations for Organic Nonlinear Materials*, Marcel Dekker Inc., 1998, pp. 671-708.

[21] S. Liu, C. Guixiang, H. Yunyu, S. Lin, Y. Zhang, M. He, W. Xiang and X. Liang, "Tunable Fluorescence and Optical Nonlinearities of All Inorganic Colloidal Cesium Lead Halide Perovskite Nanocrystals," *Journal of Alloys and Compounds,* vol. 724, pp. 889-896, 2017. https://doi.org/10.1016/j.jallcom.2017.06.034

[22] J. He, Y. Qu, H. Li, J. Mi and W. Ji, "Three-Photon Absorption in ZnO and ZnS Crystals," *Optics Express,* vol. 13, no. 23, pp. 9235-9247, 2005. https://doi.org/10.1364/OPEX.13.009235

[23] B. Wherrett, "Scaling Rules for Multiphoton Interband Absorption in Semiconductors," *Journal of the Optical Society of America B,* vol. 1, no. 1, pp. 67-72, 1984. https://doi.org/10.1364/JOSAB.1.000067

[24] H. Brandi and C. De Araujos, "Multiphoton Absorption Coefficients in Solids: a Universal Curve," *Journal of Physics C: Solid State Physics,* vol. 16, no. 30, p. 5929, 1983. https://doi.org/10.1088/0022-3719/16/30/022

[25] A. Manzi, Y. Tong, J. Feucht, E. Yao, L. Polavarapu, A. Urban and J. Feldmann, "Resonantly Enhanced Multiple Exciton Generation through Below-Band-Gap Multi-Photon Absorption in Perovskite Nanocrystals," *Nature Communications,* vol. 9, no. 1, p. 1518, 2018. https://doi.org/10.1038/s41467-018-03965-8

Plasma-Induced Generation of Optically Active Defects in Glasses

Christoph Gerhard

University of Applied Sciences and Arts, Faculty of Natural Sciences and Technology, Von-Ossietzky-Straße 99, 37085 Göttingen, Germany

christoph.gerhard@hawk.de

Keywords: Glasses, Plasma Treatment, Glass Defects, Optical Properties

Abstract. In this contribution, the modification of optical and electrical properties of glasses by plasma treatment is introduced. The presented method is based on pulsed dielectric barrier discharge plasmas which are operated with hydrogenous working gases where atomic and excited hydrogen species are generated within the plasma by electron impact-induced dissociation. These species initiate a notable modification of the glass network. As a result, the transmission characteristics and associated parameters of plasma treated glass are altered. The impact of plasma treatment on the chemical composition of optical glasses as well as the accompanying changes in optical properties are presented. As determined via secondary ion mass spectroscopy, oxygen is removed and hydrogen is implanted into the glass bulk material. This leads to the formation of oxygen vacancies such as E'-centers on the one hand and the generation of hydrogen centers on the other hand. Such glass modifications lead to a drastic increase in absorption and index of refraction as ascertained via UV/VIS-spectroscopy. Some potential applications of these plasma-induced effects are suggested.

Introduction

The local modification of optical and electrical properties of photonic materials is of great interest for a number of applications. The generation of (micro) structures with a defined spatial distribution in index of refraction and permittivity that vary from the matrix material allows for generating devices with different optical and electrical functionalities on the same substrate. Such systems are thus of great interest for the development and enhancement of micro opto-mechanical or electro-mechanical systems.

Quite a number of different approaches, techniques and methods for such local and spatially limited modification of glasses were developed in the last decades. Amongst others, this includes the implantation and exchange of ions [1,2], (femtosecond) laser-induced formation of photo-structural defects [3], glass modification by electron-beam irradiation [4] or protons [5] as well as different dry etching methods, e.g. physical ion etching or chemical plasma etching [6] as usually applied in lithography. The composition of a glass and its optical as well as its electrical properties can also be modified by the use of technical plasmas as shown by recent works in the last few years [7-9]. This approach was applied in the present work.

Experimental Approach

For plasma-induced modification of the chemical composition and thus optical and electrical properties of glasses, pulsed dielectric barrier discharge (DBD) plasmas were used in the present work. The discharge geometry was a so-called indirect plasma, which is also referred to as plasma jet or plasma torch. Here, the plasma is generated within a plasma source and blown onto the work piece surface by the process gas flow. Due to the principle of DBD, where at least one of the involved high voltage electrodes for plasma generation is dielectrically separated from the actual discharge gap, the electrical current is strictly limited. Hence, the plasma gas temperature is close to ambient temperature. DBD plasmas are thus also known as "cold" plasmas. In the present case, the maximum surface temperature of the treated glass surface was approximately 38°C as ascertained via infrared camera measurements [7]. Experiments were performed on different optical and technical glasses [10,11], but mainly on fused silica as reported in more detail elsewhere [12,13]. The process gas used was forming gas 90/10, i.e. a mixture of nitrogen (90%) and hydrogen (10%). Similar results as presented hereafter could also be achieved when using any other hydrogenous gas due to the following mechanism:

Within the plasma volume, atomic hydrogen is generated by dissociation which results from the collision of molecular hydrogen and free electrons according to

$$H_2 + e^- \rightarrow 2H + e^-. \quad (1)$$

For this effect, a quite low electron energy of 5.4 eV is required (For comparison, the energy needed for ionization of atomic hydrogen amounts to 13.6 eV). Atomic hydrogen stands out due to an extremely high reactivity [14]; this plasma species is thus capable of initiating a chemical reduction process,

$$SiO_2 + 2H \rightarrow SiO + 2H_2O, \quad (2)$$

where gaseous, volatile water is formed and oxygen is consequently removed from the glass network. Such removal results in the formation of optically active defects, i.e. oxygen deficiency centers. For instance, this includes unpaired electrons of the silicon atom, the so-called E'-centers, or the attachment of hydrogen at the original lattice site of an oxygen atom, i.e. a H(I) center, see Fig. 1.

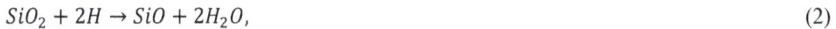

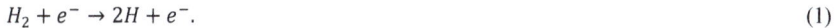

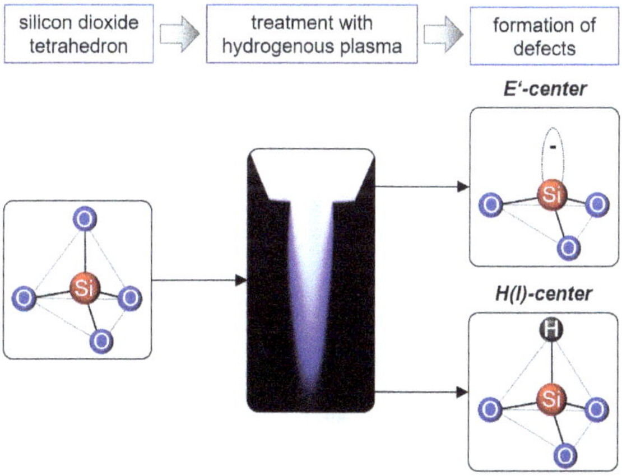

Fig. 1: Visualization of plasma-induced formation of glass defects by a modification of the network forming silicon dioxide tetrahedron (left) due to the generation of E'-centers and H(I)-centers (right).

Impact of Generated Defects on Glass Properties

It is well known that oxygen deficiency-related glass defects as introduced in the previous section feature notable optical absorption bands [15], mainly in the ultraviolet wavelength range from approximately 150-370 nm. This effect was also observed in the present work as ascertained via UV-VIS spectroscopic measurements and shown in Fig. 2.

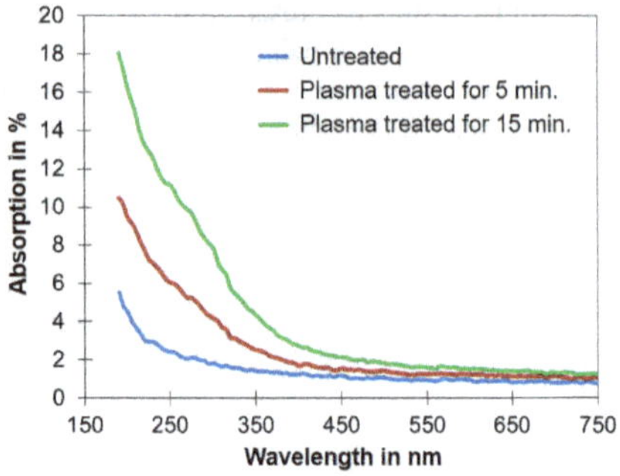

Fig. 2: Plasma-induced increase in absorption of fused silica vs. wavelength.

This result indicates the formation of UV-absorbing glass defects such as E'-centers and H(I)-centers. The removal of oxygen from the glass network, which is necessary for such defects, as well as the accompanying implantation of hydrogen was confirmed via secondary ion mass spectroscopy (SIMS) [8]. As shown in Fig. 3, oxygen was continuously removed from fused silica in the course of plasma treatment. Starting at a value of approximately 66%, the oxygen content was reduced to 55.44% after 30 minutes of plasma treatment. Simultaneously, the hydrogen content was increased from initially 0.88% to 5.15%.

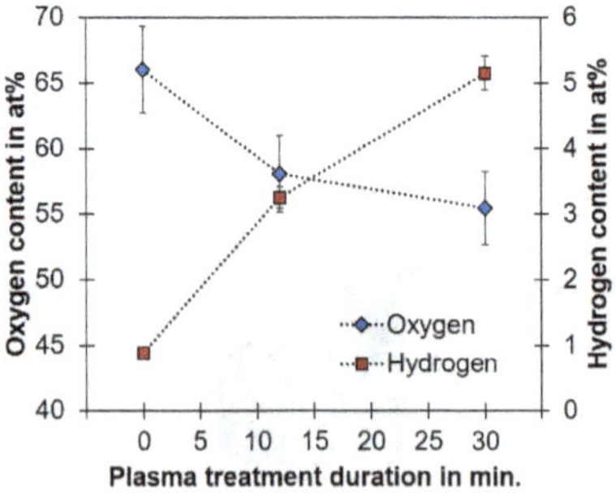

Fig. 3: Oxygen and hydrogen content in fused silica vs. plasma treatment duration.

Since the silicon content remains constant at approximately 33%, the stoichiometry factor x which represents the ratio of oxygen and silicon, O/Si, decreases over plasma treatment duration as shown in Figure 4. After 30 minutes, it amounts to 1.68. Since O/Si is 2 in the case of pure fused silica or SiO_2, it turns out that silicon suboxide SiO_x is generated by the plasma treatment. This notable modification of the chemical composition of fused silica results in the increase in absorption shown in Fig. 2 and an increase in the extinction coefficient k, i.e. the imaginary part of the complex index of refraction, respectively. Consequently, the index of refraction n, which is the real part of the complex index of refraction, increases as well as ascertained via ellipsometric and interferometric measurements [16], see Fig. 5. Since the index of refraction of a transparent medium is directly related to its electrical permittivity ε ($n \approx \sqrt{\varepsilon}$), the latter parameter is modified as well by the plasma-induced formation of silicon suboxide. As an approach, the permittivity ε of SiO_x can be estimated on the basis of the particular permittivities of pure silicon and silicon dioxide according to Philipp's interpolation formula [17], given by

$$\varepsilon_{SiO_x} \approx \left(1 - \frac{x}{2}\right)^2 \cdot \varepsilon_{Si} + \left(\frac{x}{2}\right) \cdot \varepsilon_{SiO_2}. \tag{3}$$

When applying this equation to the data for the stoichiometry factor x obtained from SIMS measurements it turns out that the permittivity increases in the course of the plasma treatment due to a decrease in x as visualized in Fig. 4. Here, the particular permittivities at a wavelength of 1.5 µm, i.e. $\varepsilon_{SiO} = 11.9$ F/m and $\varepsilon_{SiO2} = 2.1$ F/m were used for calculation.

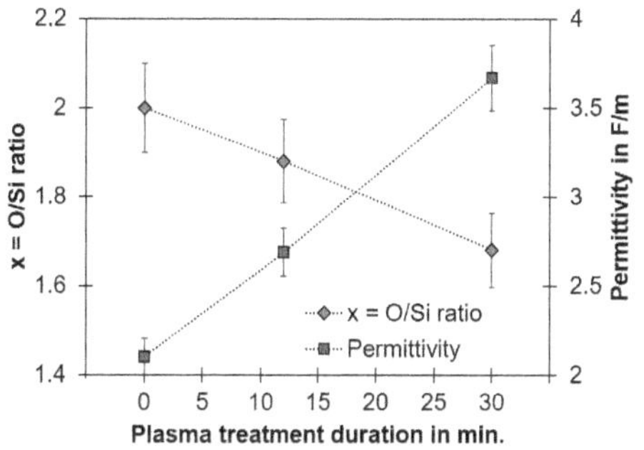

Fig. 4: Stoichiometry factor x = O/Si and permittivity of fused silica vs. plasma treatment duration.

It can finally be stated that the applied plasma treatment leads to a modification of the chemical composition of fused silica as discussed in more detail in [8,9]. Consequently, the optical and electrical properties, i.e. the absorption, index of refraction and permittivity are altered.

Applications

The above-described plasma-induced mechanisms and effects have a number of potential applications in optics and photonics. For instance, a local point-shaped treatment using a (micro) plasma source could be used in order to initiate a three-dimensional chemically modified glass volume within undistorted fused silica bulk material. This approach is quite comparable to an ion exchange process as applied for the generation of planar gradient index (GRIN) lenses [18]. Hence, experiments were performed on a plane fused silica sample using a plasma jet source with an effective plasma diameter of approximately 2 mm. Before and after treatment, the wave front distortion of a collimated laser beam at a wavelength of 405 nm was measured in order to identify the impact of the plasma-modified glass volume on the laser irradiation. Based on this measurement, the change in focal length of the sample was determined as shown in Fig. 5. It turns out that a slightly focusing element was generated within the glass volume due to an increase in index of refraction. However, this effect is quite negligible due to the extremely low depth of penetration of plasma-generated effects which – at a maximum – amounts to some microns [8]. Here, further improvements of the process by changing the plasma parameters could provide a novel approach for the realization of embedded optical components in the future.

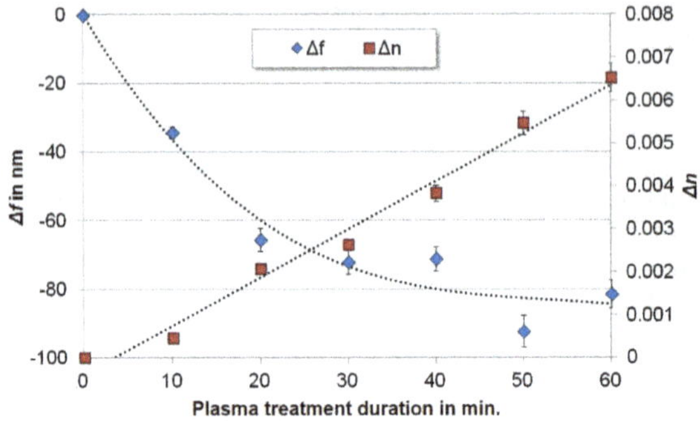

Fig. 5: Change in focal length Δf of a fused silica sample as well as its change in index of refraction Δn (both values given for a wavelength of 405 nm) vs. plasma treatment duration.

Another point is the plasma-induced increase in absorption in ultraviolet wavelength range. This effect represents an advantage for laser-based micro structuring of transparent media. In order to overcome the high transparency of glasses for nearly all commercially available and established laser sources, a number of approaches and techniques were developed in the past. Most of these techniques are based on the application of absorbing layers such as toluene (C_7H_8), pyrene ($C_{16}H_{10}$), or silicon monoxide (SiO) to the glass surface. This involves laser-induced backside wet etching (LIBWE) [19], laser etching at surface adsorbed layers (LESAL) [20,21], and laser-induced backside dry etching (LIBDE) [22].

In the present case, the plasma-modified near-surface glass material consisting of sub-stoichiometric silicon oxide and implanted hydrogen – and thus featuring increased absorption – acts as a solid absorbing layer where the absorption depends on wavelength, plasma treatment duration and plasma power [23]. At the beginning of a plasma treatment process the increase in absorption, i.e. the decrease in transmission per wavelength as determined via UV/VIS-spectroscopy increases linearly over time. As shown in Fig. 6, the efficiency is higher for shorter wavelengths.

Laser ablation with laser sources in the ultraviolet wavelength range is thus improved by the plasma-induced increase in absorption and the resulting enhanced coupling of incoming laser energy into the glass surface. Experiments applying an ArF-excimer laser with an emission wavelength of 193 nm have shown that after plasma pre-treatment for 15 minutes, the energy required for obtaining laser-induced material removal can be reduced by a factor of 4.6. As a result, the contour accuracy and surface of laser-generated structures are significantly improved [12,24]. Moreover, the ablation rate and thus the material removal efficiency is increased [25].

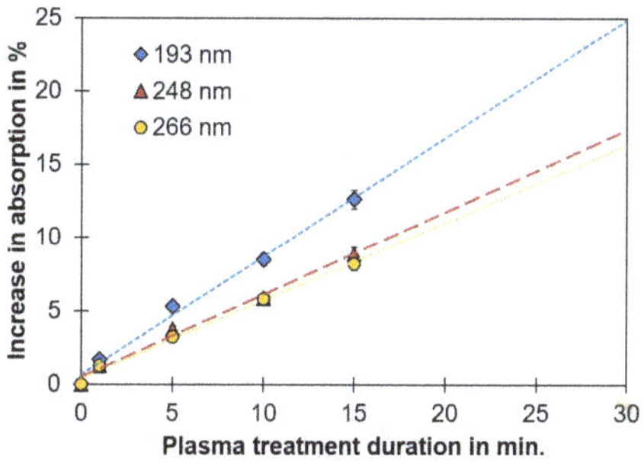

Fig. 6: Increase in absorption for three different laser wavelengths (linear regression fits of measured data are displayed) vs. plasma treatment duration.

Summary and Outlook

By applying the presented approach of plasma-induced modification of the chemical composition, the optical and electrical properties of glasses are altered. The introduced method thus represents a potential tool for different applications in photonics. For instance, the use of micro plasmas allows for the generation of structures with modified optical and electrical properties within an unmodified glass matrix, e.g. for structuring of waveguides.

References

[1] M. Oikawa, K. Iga, T. Sanada, N. Yamamoto, K. Nishizawa, Array of distributed-index planar micro-lenses prepared from ion exchange technique, Jap. J. Appl. Phys. 20 (1981) L296-298. https://doi.org/10.1143/JJAP.20.L296

[2] J. Qiu, X. Jiang, C. Zhu, H. Inouye, J. Si, K. Hirao, Optical properties of structurally modified glasses doped with gold ions, Opt. Lett. 29 (2004) 370-372. https://doi.org/10.1364/OL.29.000370

[3] A. Zoubir, C. Rivero, R. Grodsky, K. Richardson, M. Richardson, T. Cardinal, M. Couzi, Laser-induced defects in fused silica by femtosecond IR irradiation, Phys. Rev. B 73 (2006) 224117. https://doi.org/10.1103/PhysRevB.73.224117

[4] N. Jiang, J. Qiu, A.L. Gaeta, J. Silcox, Nanoscale modification of optical properties in Ge-doped SiO2 glass by electron-beam irradiation, Appl. Phys. Lett. 80 (2002) 2005-2007. https://doi.org/10.1063/1.1454211

[5] M. Frank, M. Kufner, S. Kufner, M. Testorf, Microlenses in polymethyl methacrylate with high relative aperture, Appl. Opt. 30 (1991) 2666-2667. https://doi.org/10.1364/AO.30.002666

[6] C. Gerhard, Optics Manufacturing: Components and Systems, 1rst ed., CRC Taylor & Francis, Boca Raton, 2017. https://doi.org/10.1201/9781351228367

[7] A. Gredner, C. Gerhard, S. Wieneke, K. Schmidt, W. Viöl, Increase in generation of poly-crystalline silicon by atmospheric pressure plasma-assisted excimer laser annealing, Journal of Mater. Sci. Eng. B 3 (2013) 346-351. https://doi.org/10.17265/2161-6221/2013.06.002

[8] C. Gerhard, D. Tasche, S. Brückner, S. Wieneke, W. Viöl, Near-surface modification of optical properties of fused silica by low-temperature hydrogenous atmospheric pressure plasma, Opt. Lett. 37 (2012) 566-568. https://doi.org/10.1364/OL.37.000566

[9] C. Gerhard, T. Weihs, D. Tasche, S. Brückner, S. Wieneke, W. Viöl, Atmospheric pressure plasma treatment of fused silica, related surface and near-surface effects and applications, Plasma Chem. Plasma P. 33 (2013) 895-905. https://doi.org/10.1007/s11090-013-9471-7

[10] C. Gerhard, J. Heine, S. Brückner, S. Wieneke, W. Viöl, A hybrid laser-plasma ablation method for improved nanosecond laser machining of heavy flint glass, Laser Eng. 24 (2013) 391-403.

[11] C. Gerhard, M. Dammann, S. Wieneke, W. Viöl, Sequential atmospheric pressure plasma-assisted laser ablation of photovoltaic cover glass for improved contour accuracy, Micromachines 5 (2014) 408-419. https://doi.org/10.3390/mi5030408

[12] J. Hoffmeister, C. Gerhard, S. Brückner, J. Ihlemann, S. Wieneke, W. Viöl, Laser micro-structuring of fused silica subsequent to plasma-induced silicon suboxide generation and hydrogen implantation, Physics Proc. 39 (2012) 613-620. https://doi.org/10.1016/j.phpro.2012.10.080

[13] C. Gerhard, J. Heine, S. Brückner, S. Wieneke, W. Viöl, A hybrid laser-plasma ablation method for improved nanosecond laser machining of heavy flint glass, Laser Eng. 24 (2013) 391-403.

[14] H. Shirai, Y. Fujimura, S. Jung S, Formation of nanocrystalline silicon dots from chlorinated materials by RF plasma-enhanced chemical vapor deposition, Thin Solid Films 407 (2002) 12-17. https://doi.org/10.1016/S0040-6090(02)00005-6

[15] L. Skuja, Optically active oxygen-deficiency-related centers in amorphous silicon dioxide, J. Non-Cryst. Solids 239 (1998) 16-48. https://doi.org/10.1016/S0022-3093(98)00720-0

[16] C. Gerhard, M. Kretschmer, W. Viöl, Plasma meets glass - plasma-based modification and ablation of optical glasses, Optik & Photonik 7 (2012) 35-38. https://doi.org/10.1002/opph.201290098

[17] H. R. Philipp, Optical properties of non-crystalline Si, SiO, SiO_x and SiO_2, J. Phys. Chem. Solids 32 (1971) 1935-1945. https://doi.org/10.1016/S0022-3697(71)80159-2

[18] H. Ottevaere, R. Cox, H.P. Herzig, T. Miyashita, K. Naessens, M. Taghizadeh, R. Völkel, H.J. Woo, H. Thienpont, Comparing glass and plastic refractive microlenses fabricated with different technologies, J. Opt. A: Pure Appl. Opt. 8 (2006) S407-S429. https://doi.org/10.1088/1464-4258/8/7/S18

[19] J. Wang, H. Niino, A. Yabe, One-step microfabrication of fused silica by laser ablation of an organic solution, Appl. Phys. A 68 (1999) 111-113. https://doi.org/10.1007/s003390050863

[20] R. Böhme, K. Zimmer, Low roughness laser etching of fused silica using an adsorbed layer, Appl. Surf. Sci. 239 (2004) 109-116. https://doi.org/10.1016/j.apsusc.2004.05.095

[21] K. Zimmer, R. Böhme, B. Rauschenbach, Laser etching of fused silica using an adsorbed toluene layer, Appl. Phys. A 79 (2004) 1883-1885. https://doi.org/10.1007/s00339-004-2961-y

[22] B. Hopp, C. Vass, T. Smausz, Z. Bor, Production of submicrometre fused silica gratings using laser-induced backside dry etching technique, J. Phys. D 39 (2006) 4843-4847. https://doi.org/10.1088/0022-3727/39/22/015

[23] C. Gerhard, E. Letien, T. Cressent, M. Hofmann, Impact of the plasma power on plasma-induced increase in absorption of fused silica, Wiss. Beitr. 23 (2019) 33-37.

[24] S. Brückner, J. Hoffmeister, J. Ihlemann, C. Gerhard, S. Wieneke, W. Viöl, Hybrid laser-plasma micro-structuring of fused silica based on surface reduction by a low-temperature atmospheric pressure plasma, J. Laser Micro Nanoen. 7 (2012) 73-76. https://doi.org/10.2961/jlmn.2012.01.0014

[25] D. Tasche, C. Gerhard, J. Ihlemann, S. Wieneke, W. Viöl, The impact of O/Si ratio and hydrogen content on ArF excimer laser ablation of fused silica, J. Eur. Opt. Soc-Rapid 9 (2014) 14026 (4pp). https://doi.org/10.2971/jeos.2014.14026

Preliminary Data on a SERS-Responsive Sensor Based on Metallic Nanostructures Functionalized by Aptamers Specific for Arsenic

Domenica Musumeci[1,a *], Daniela Montesarchio[1,b], Elisa Scatena[2,c], Costantino Del Gaudio[2,d], Fabio De Matteis[3,e], Roberto Francini[3,f], Mauro Casalboni[3,g *]

[1] Dept. of Chemical Sciences, Federico II University of Napoli, via Cinthia 4, 80126 Napoli, Italy

[2] Fondazione E. Amaldi, via del Politecnico snc, 00133 Roma, Italy

[3] Dept. of Industrial Engineering, Tor Vergata University of Roma, via del Politecnico 1, 00133 Roma, Italy

[a]domenica.musumeci@unina.it, [b]daniela.montesarchio@unina.it, [c]elisa.scatena@fondazioneamaldi.it, [d]costantino.delgaudio@fondazioneamaldi.it, [e]dematteis@roma2.infn.it, [f]francini@roma2.infn.it, [g]casalboni@uniroma2.it

* corresponding authors

Keywords: Arsenic Detection, SERS, Oligonucleotide Aptamers

Abstract. Arsenic, in the form of arsenate (As^{V}) and arsenite (As^{III}), is a toxic carcinogen widely distributed in aqueous environments in many parts of the world. Efficient arsenic sensors in terms of sensitivity, selectivity, speed and portability are urgently needed. The present research was focused on the development of a gold nanostructured sensor, functionalized with organic molecules able to selectively bind arsenic, for SERS detection. As organic molecule, we here selected oligonucleotide aptamers specific for arsenic recognition relying on previous studies on a 100-mer arsenic-binding DNA aptamer (Ars3), selected for its highest affinity and specificity to arsenate and arsenite. The aptamer Ars3 was previously used as such in its whole sequence and no attempt to optimize it, in terms of size and sensor efficiency, or to unravel its binding mechanism with arsenic has been carried out. Furthermore, even if in the previously proposed Ars3-based sensors a low detection limit for As was achieved, the assembly of the probe and the detection methodologies were in most cases very complex and not suitable for the development of portable in situ devices. Thus, in order to optimize Ars3 and investigate its interaction with arsenic, we here designed shorter DNA sequences cutting the 3' and 5' ends of the parent aptamer, and carried out spectroscopic and electrophoretic analysis together with arsenic-binding assays by using a suitably functionalized affinity resin. In addition, a specific SERS-responsive system with the Ars3 parent aptamer was considered in view of its application with the modified DNA sequences here proposed. Collected findings highlighted that the parent aptamer did not bind arsenic with high affinity. This was also in agreement with very recent results, published concomitantly with our studies, which stated that Ars3 was not able to bind As and that all the positive binding data using Ars3 were due to the use of gold supports (SPR chips or gold nanoparticles) which tightly bound arsenic. Based on these results, no arsenic-binding DNA aptamers are currently known, thus underlining the need for actual arsenic binding aptamers to be implemented for SERS sensors.

Introduction

Arsenic poisoning is caused by elevated levels of arsenic in the body. The dominant basis of arsenic poisoning is from ground water that naturally contains high concentration of arsenic [1]. Chronic ingestion or long-term exposure to arsenic can cause a variety of cancer forms, skin lesions, and cardiovascular diseases [1]. Elevated concentrations of arsenic in ground water have been found in many parts of the world and the maximum tolerable value of arsenic in drinkable water, in agreement with the World Health Organization, is 10 μg/L (ppb) [2]. It is generally recognized that inorganic arsenic is more toxic than organic one, and among inorganic arsenic, arsenic in oxidation state +3 (i.e., arsenite, derived from arsenous acid) is more toxic than that in the oxidation state +5 (i.e., arsenate, derived from arsenic acid) [1].

Currently, the most used analytical methodologies for the detection of arsenic at very low concentrations are Atomic Fluorescence Spectroscopy (AFS), Atomic Absorption Spectroscopy (AAS) and Inductively Coupled Plasma Mass Spectrometry (ICP-MS) [1],[3],[4]. These techniques are very efficient, but not suitable for *in situ* analysis, expensive, time consuming, and require chemical and physical pre-treatment of the samples.

Therefore, it is essential not only to develop sensitive and selective analytical techniques for the detection of arsenic even in traces in the environment, but also to construct efficient arsenic-specific sensors in terms of speed and portability, especially for developing countries with high-risk contaminated soils. In this context, Surface-Enhanced Raman Scattering (SERS) guarantees high sensitivity and allows engineering tools of extremely reduced size. SERS can accomplish the speciation analysis for species with different oxidation states, with no need for any complex preparation of the sample, and has thus emerged as an extremely promising solution for *in situ* detection of arsenic in the field, particularly when coupled with portable/handheld Raman spectrometers [1],[5].

The present research was focused on the realization of a gold nanostructured sensor for SERS detection, functionalized with organic molecules able to selectively bind arsenic. As organic molecules, we here selected oligonucleotide aptamers specific for arsenic recognition.

Aptamers are DNA/RNA oligonucleotides (ON) able to recognize with high affinity and selectivity a specific target based on their peculiar three-dimensional structures [6]. Aptamers can be selected from huge libraries of molecules containing randomly generated sequences, and specifically bind different targets, from proteins, viruses or whole cells to small molecules or ions [6],[7].

About 10 years ago a Korean group identified a 100-mer DNA aptamer (named Ars3) having high affinity and specificity for arsenate (As^{V}) and arsenite (As^{III}), with dissociation constants in the low nM range [8]. This arsenic-binding aptamer was selected from a random DNA library of 100-mer oligonucleotides which contained two constant regions at their ends as well as a 40 nt random region in the middle: $^{5'}$GGT AAT ACG ACT CAC TAT AGG GAG ATA CCA GCT TAT TCA ATT-**N**$_{40}$-AGA TAG TAA GTG CAA TCT$^{3'}$.

The aptamers were selected using affinity chromatography-based SELEX (Systematic Evolution of Ligands by EXponential enrichment), obtained by immobilizing arsenic on an agarose resin. Quantitative analyses of the aptamer candidates by SPR revealed Ars3 as the aptamer with the highest affinity to arsenate and arsenite. The specific affinity interactions of the aptamer to arsenic were verified against other heavy metals as negative controls, confirming its high specificity [8].

From this first work, Ars3 has been explored also in subsequent researches on arsenic sensors. In all cases, the aptamer was used as such, that is containing the random part plus the two flanking constant regions (which were generally removed) required for the PCR (polymerase chain reaction) step of the SELEX, and no attempt to optimize it, in terms of size and sensor-efficiency, or to elucidate its binding mechanism with arsenic has been ever described [9-17]. Furthermore, even if the detection limit for As sensors previously reported exploiting Ars3 was generally low, the assembly of the probe and the detection methodologies were very complex and not suitable for the development of portable in situ devices.

Thus, in order to optimize Ars3 and further analyse its interaction with arsenic, we designed shorter DNA sequences, cutting in various way the 3' and 5' ends of the parent aptamer, and carried out spectroscopic and electrophoretic analyses together with arsenic-binding assays by using a suitably functionalized affinity resin. Moreover, a SERS-responsive system to Ars3 was ad hoc implemented in view of its application with the modified DNA sequences here proposed.

Results and Discussion

The detection procedure and the arsenic-binding aptamer Ars3 optimization were both addressed, designing a set of shorter oligonucleotide sequences starting from the original 100-mer, cutting its constant parts in various ways but leaving unaltered the central core of the sequence, which should contain the residues necessary for the specific recognition.

From the analysis of all the designed sequences using *m-fold* program we selected three sequences (**Ars40**, **Ars58** e **Ars64**, Fig. 1) potentially able to form hairpin-loop structures with good stability as the parent aptamer.

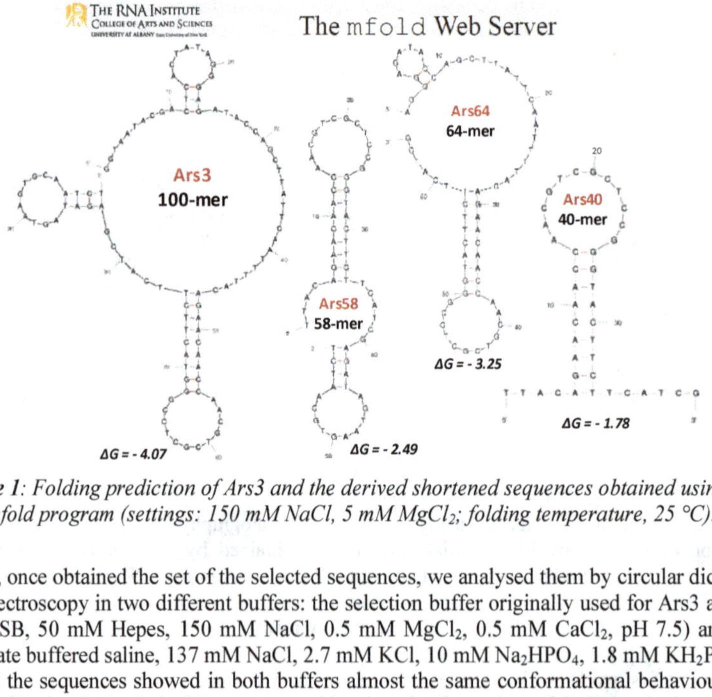

***Figure 1**: Folding prediction of Ars3 and the derived shortened sequences obtained using the mfold program (settings: 150 mM NaCl, 5 mM $MgCl_2$; folding temperature, 25 °C).*

Then, once obtained the set of the selected sequences, we analysed them by circular dichroism (CD) spectroscopy in two different buffers: the selection buffer originally used for Ars3 aptamer (named SB, 50 mM Hepes, 150 mM NaCl, 0.5 mM $MgCl_2$, 0.5 mM $CaCl_2$, pH 7.5) and PBS (phosphate buffered saline, 137 mM NaCl, 2.7 mM KCl, 10 mM Na_2HPO_4, 1.8 mM KH_2PO_4, pH 7.4). All the sequences showed in both buffers almost the same conformational behaviour, with quite similar bands in their CD spectra, evidencing the formation of B-DNA duplex tracts – probably due to hairpin-loop structures (Fig. 2a). CD-melting curves showed the highest melting temperature (T_m) for the 100-mer ON and the lowest T_m value for the 40-mer (Fig. 2b). All the sequences were analysed in both annealed (A) and not annealed (NA) form in order to explore both the thermodynamically and kinetically favoured conformations, respectively. The CD spectra of both the annealed and not-annealed samples were in all cases quite superimposable. This result proved that the kinetically more stable conformations were essentially similar to the thermodynamically favoured ones for each ON.

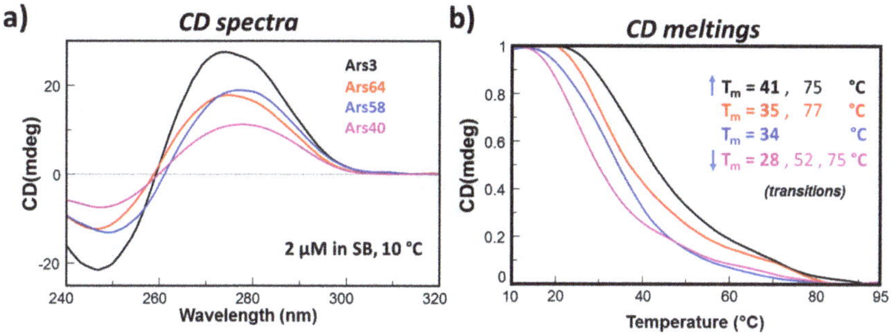

***Figure 2**: a) Overlapped CD spectra of the four not-annealed ONs in SB at 2 µM conc. at 10 °C; b) normalized CD-melting curves of the solutions in a).*

Analysis of the electrophoretic mobility of the annealed and not annealed forms of the four investigated sequences through native PolyAcrilamide Gel Electrophoresis (PAGE) excluded the formation of bimolecular or higher-order structures since a single band for each oligonucleotide was observed. Thus, all the sequences formed exclusively monomolecular species in the selected buffer and concentration conditions (Fig. 3).

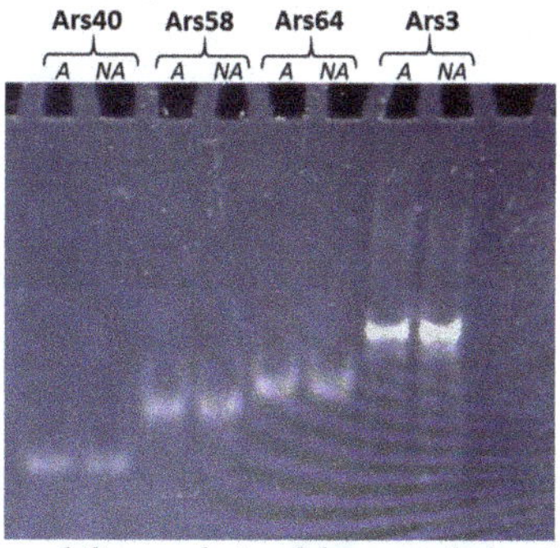

***Figure 3**: Electrophoretic mobility analysis of the annealed and not annealed forms of the investigated sequences at 2 µM conc. in SB. Native 7 % polyacrylamide gel electrophoresis (PAGE) with TBE 1x as running buffer.*

In parallel, in order to design the arsenic sensor and set up the system for arsenic detection, we preliminarily attached the parent aptamer (that is the 100-mer Ars3) on a substrate suitable for SERS analysis. This substrate was based on highly packed gold-coated nanopillars structured on a silicon wafer. The aptamer was modified at its 5' end with a C6-linker carrying a thiol group useful for the immobilization on the nanopillars *via* Au-S covalent bonds (Fig. 4). In brief, after the deposition of the target solution on the functionalized substrate, the wetted surface was allowed to dry. Due to the surface tension between pillars during liquid evaporation, the nanopillars, leaning toward each other, formed micro-sized clusters with highly localized plasmonic field around the contact region (the so-called hotspots). When the analyte was trapped by an aptamer within a hotspot, the Raman signal was highly enhanced.

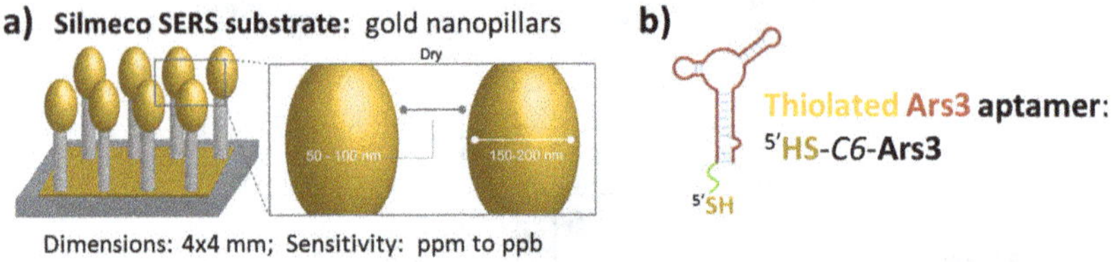

Figure 4: Schematic representations of a) a Silmeco substrate for SERS analysis and b) the 5'-thiolated Ars3 aptamer.

Thus, we prepared two substrates: 1) a reference substrate containing Ars3 and the 6-mercapto-1-hexanol (MCH) as passivating agent, and 2) another substrate containing also As^{III} added as analyte. In a preliminary assay, we analysed in label-free mode the two prepared substrates and evidenced some remarkable differences in their SERS spectra (Fig. 5).

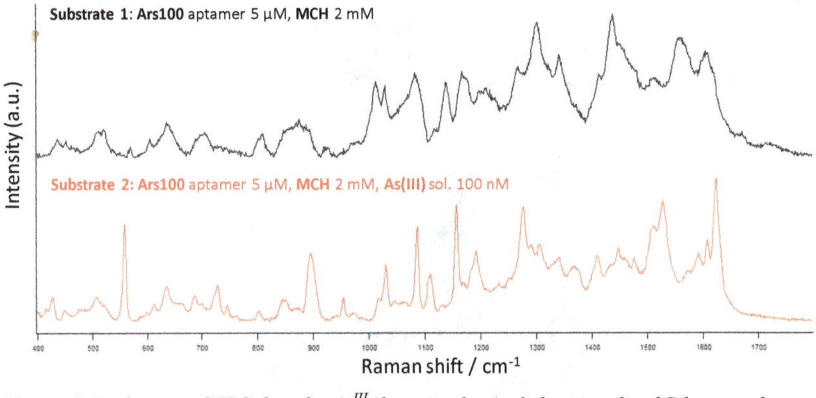

Figure 5: Preliminary SERS data for As^{III} detection by Ars3-functionalized Silmeco substrates.

We then performed CD-titrations in SB or PBS solutions of the original aptamer and the shorter analysed sequences (in both annealed and not-annealed form) with both As^{III} and As^{V} salts in order to investigate the possible conformational changes due to arsenic binding and thus measure the affinity and stoichiometry of these interactions.

In all cases the CD spectra of all the tested ONs remained unchanged upon As titration (data not shown): thus, even if the binding with arsenite and arsenate occurred, this did not affect the conformation of the ONs, including the parent aptamer Ars3. This result was in net disagreement with various works reporting that marked CD changes occurred after arsenic addition to Ars3 [9],[11],[15],[16].

Materials Research Forum LLC
https://doi.org/10.21741/9781644900710-6

Furthermore, also the thermodynamic stability of the structured ONs (monitored through CD-melting experiments) did not show variations in the presence of arsenic: indeed, the melting temperatures (T_m) for all the investigated systems (ON-As) remained almost unchanged, at least in the selected concentration and buffer conditions, with respect to the solutions of the free ONs. Therefore, from CD experiments we were not able to measure the affinity of Ars3 and of the other sequences for arsenic since no change in the CD spectra derived from arsenic addition to the ONs was evidenced: this behaviour however was not unusual since it was found also for other DNA binders, especially those providing groove interactions with duplex or G-quadruplex DNA structures.

Subsequently, in order to further check the binding affinity of Ars3 for arsenic, using a technique alternative to SPR or CD, we decided to exploit affinity chromatography taking inspiration from the first work on Ars3. We thus anchored arsenic(III) (in the form of its salt $NaAsO_2$) on an agarose resin (Fig. 6a). First, the resin, containing activated carboxylic groups was covalently coupled with 1,8-diaminooctane and its incorporation was qualitatively evaluated through ninhydrin test (few mg of resin containing free amino groups, assumed a strong blue-violet colour reacting at high temperature with ninhydrin; on the contrary the control amine-free resin remained colourless, Fig. 6b). Then, arsenite was linked exploiting electrostatic interactions between its negative charge and the ammonium moiety on the resin, using a protocol optimized for the pH and concentration of the arsenite solution (Fig. 10a).

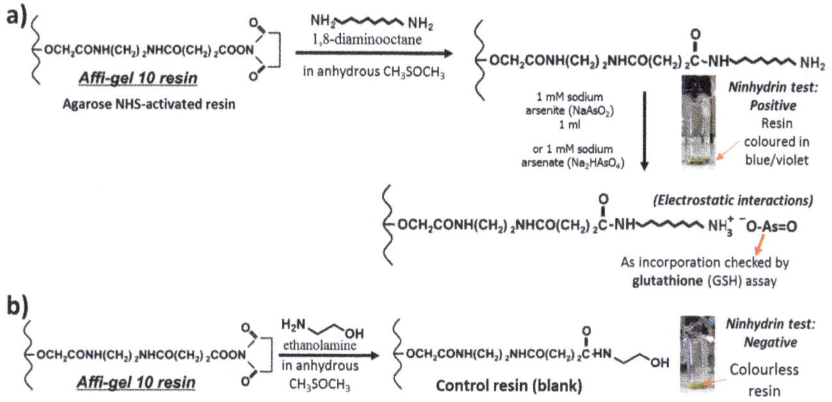

***Figure 6**: Schematic representation of the preparation of the As-funcionalized (a) and control (b) resins.*

Arsenic incorporation on the resin was evaluated through glutathione (GSH) assay, *ad hoc* adapted for the system we previously assembled in solid phase. This assay is based on the formation of a stable complex between the thiol group of the GSH and $arsenic^{III}$ which shows a characteristic UV-band at 290 nm [18].

Thus, after incubation of the amine resin with arsenite solution, first, the excess of arsenic was washed away, and then the As^{III} bound to the resin was subsequently released using a 20 mM GSH solution and quantified by UV (Fig. 7).

Subsequently, we planned to set up the As^{III} affinity assay for Ars3 and the modified sequences. The solutions of the four oligonucleotides (both in annealed and not-annealed form) in PBS or SB were loaded on the resin separately and the eluted fractions were collected, including various washings, and analysed by UV measurements. From the difference in the absorption of all the ON solutions (band at 260 nm) before and after loading them on the resin, we calculated the percentage content of ON captured by the resin and thus the affinity of arsenic for each sequence, including the parent Ars3 aptamer. The unspecific absorption on the blank resin, which however was minimal, was also considered and subtracted.

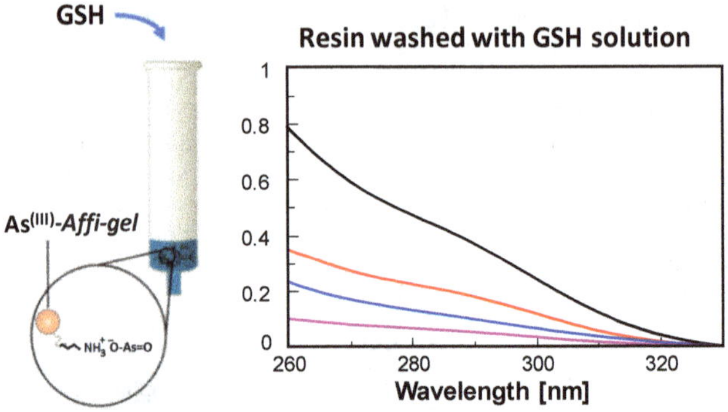

***Figure** 7: Schematic representation of the GSH assay on the As-functionalized Affigel resin and overlapped UV spectra of the fraction eluted from the resin after GSH solution washings.*

Unfortunately, the binding assay did not give the expected results (neither for the reference aptamer Ars3) and always provided not reproducible data. Thus, after a number of experiments, performed changing various parameters of the binding conditions (concentrations, temperature, incubation time, etc.), we concluded that Ars3 did not show the expected nM affinity for arsenite, and that As^{III} was washed away from the resin in various amount together with the oligonucleotide solutions or the buffer, as verified by the GSH assay.

Probably, the electrostatic interactions were not sufficient to keep the arsenite stably linked on the resin as stated in ref. [8] relatively to the SPR binding assays. Thus, we found many contradictory results with respect to the first work on the arsenic binding aptamer Ars3 [8]. Indeed, in parallel to our experiments two very recent works demonstrated the high affinity of arsenite for gold and the false positive generated from the interactions of aptamer and As^{III} on gold surface-involving assays [19-20]. Particularly, in ref [20], the ability of aptamer Ars3 to bind arsenite and arsenate was confuted by accurate ITC (Isothermal Titration Calorimetry) measurements. Thus, our difficulty to reproduce the binding assay of Ars3 with As^{III} on an agarose affinity resin (instead of the original gold-based SPR assay) was actually due to the absence of affinity of the aptamer for the arsenic target, as well as to the partial detachment from the resin of electrostatically-immobilized arsenic after repeated washings of the resin with buffered solutions.

Thus, on the basis of the recent literature reports [19-20] and of our scientific results, we can definitely exclude the DNA aptamer Ars3 as a candidate for specifically bind arsenic in a SERS-based sensor. Our studies are now directed towards peptide aptamers, specifically focussing on high cysteine-content peptides developed *ad hoc* for arsenic(III) sensing and using silver substrates for SERS detection instead of gold.

Experimental part

The DNA sequences used in this study are:

Ars-3:$^{5'}$GGTAATACGACTCACTATAGGGAGATACCAGCTTATTCAATT**TTACAGAACA
ACCAACGTCGCTCCGGGTACTTCTTCATCG**AGATAGTAAGTGCAATCT$^{3'}$(100mer).

Ars64: $^{5'}$AGGGAGATACCAGCTTATTCAATT**TTACAGAACAACCAACGTCGCTCC
GGGTACTTCTTCATCG**$^{3'}$ (64-mer, excluding 18 bases both at 5' and 3' ends);

Ars58: $^{5'}$**TTACAGAACAACCAACGTCGCTCCGGGTACTTCTTCATCG**AGATAGT
AAGT GCAATCT$^{3'}$ (58-mer, excluding the 42-base constant sequence at 5' end);

Ars40: $^{5'}$**TTACAGAACAACCAACGTCGCTCCGGGTACTTCTTCATCG**$^{3'}$ (40 mer, excluding both the constant regions of 42 and 18 bases at 5' and 3' ends, respectively).

All the ON sequences were purchased by biomers.net GmbH in lyophilized and desalted form. The annealed (A) aptamer solutions were prepared by diluting the stock aptamer solution in the selected buffer, heating it for 5 min at 95 °C and then allowing it to slowly cool to r.t. In turn, the not annealed (NA) aptamer solutions were prepared by diluting the stock aptamer solution in H_2O at r.t. in the selected buffer. CD melting experiments were performed by following the CD-band maximum of each ON upon increasing the temperature (scan rate: 1 °C/min).

Electrophoresis analysis: run, 89 V 1 h,30 min; staining, Gel green; visualization, UV ChemiDoc transilluminator.

SERS analysis: excitation source, 785 nm He-Ne laser; grating, 1200; objective: 50x; exp. Time 10s.

CD titration: ON solutions (2 μM in A and NA form) in PBS and SB were titrated with $NaAsO_2$ and $NaAsO_4$ (conc. from 2 μM to 1 mM) and CD spectra were registered at different time periods (3 min, 30 min, 2 h, 24 h).

Summary

The present research was focused on the development of a gold nanostructured sensor, functionalized with DNA aptamers able to selectively bind arsenic, for SERS detection. In order to optimize the previously reported aptamer Ars3 (a 100-mer DNA) and investigate its interaction with arsenic, we here designed shorter DNA sequences cutting the 3' and 5' ends of the parent aptamer and carried out spectroscopic and electrophoretic analysis together with arsenic-binding assays by using a suitably functionalized affinity resin. In addition, a specific SERS-responsive system with the Ars3 parent aptamer was considered in view of its application with the modified DNA sequences here proposed. Collected findings highlighted that the parent aptamer did not bind arsenic with high affinity, in agreement with very recent results, published concomitantly with our studies, which stated that Ars3 was not able to bind As. Based on these results, no DNA aptamers specifically recognizing arsenite/arsenate are currently known, thus underlining the need for actual arsenic binding aptamers to be implemented for SERS sensors.

Acknowledgements

This research was funded by Regione Lazio, through Progetto di ricerca 85-2017-15173 according to L.R. 13-08. Progetto NARAS.

References

[1] M. Bissen, F.H. Frimmel, Arsenic: A review. Part I: Occurrence, toxicity, speciation, mobility, Acta. Hydrochim. Hydrobiol. 31 (2003) 9-18. https://doi.org/10.1002/aheh.200390025

[2] Information on https://www.who.int/news-room/fact-sheets/detail/arsenic

[3] N. Yogarajah, S.S.H. Tsai, Detection of trace arsenic in drinking water: challenges and opportunities for microfluidics, Environ. Sci.: Water Res. Technol. 1 (2015) 426-447. https://doi.org/10.1039/C5EW00099H

[4] A.K Farzana, Z. Chen, L. Smith, D. Davey, R. Naidu, Speciation of arsenic in ground water samples: A comparative study of CE-UV, HG-AAS and LC-ICP-MS, Talanta 68 (2005) 406-15. https://doi.org/10.1016/j.talanta.2005.09.011

[5] J. Hao, M.J. Han, S. Han, X. Meng, T.L. Su, Q.K. Wang, SERS detection of arsenic in water: A review, J. Environ. Sci. (China) 36 (2015) 152-62. https://doi.org/10.1016/j.jes.2015.05.013

[6] H. Kaur, J.G. Bruno, A. Kumar, T.K. Sharma, Aptamers in the Therapeutics and Diagnostics Pipelines, Theranostics, 8 (2018) 4016-4032. https://doi.org/10.7150/thno.25958

[7] C. Platella, C. Riccardi, D. Montesarchio, G.N. Roviello, D. Musumeci, G-quadruplex-based aptamers against protein targets in therapy and diagnostics, Biochim. Biophys. Acta Gen. Subj. 2017;1861(5 Pt B):1429-1447. https://doi.org/10.1016/j.bbagen.2016.11.027

[8] Kim M, Um HJ, Bang S, Lee SH, Oh SJ, Han JH, Kim KW, Min J and Kim YH. Arsenic Removal from Vietnamese Groundwater Using the Arsenic-Binding DNA Aptamer. *Environ. Sci. Technol.* 2009, ***43***: 9335-9340. https://doi.org/10.1021/es902407g

[9] Wu Y, Liu L, Zhan S, Wang F, Zhou P. Ultrasensitive aptamer biosensor for arsenic(III) detection in aqueous solution based on surfactant-induced aggregation of gold nanoparticles. *Analyst*. 2012, *137*: 4171-8. https://doi.org/10.1039/c2an35711a

[10] Wu Y, Zhan S, Wang F, He L, Zhi W, Zhou P. Cationic polymers and aptamers mediated aggregation of gold nanoparticles for the colorimetric detection of arsenic(III) in aqueous solution. Chem Commun (Camb). 2012 May 11;48(37):4459-61. https://doi.org/10.1039/c2cc30384a

[11] Wu Y, Zhan S, Xing H, He L, Xu L, Zhou P. Nanoparticles assembled by aptamers and crystal violet for arsenic(III) detection in aqueous solution based on a resonance Rayleigh scattering spectral assay. Nanoscale. 2012 Nov 7;4(21):6841-9. https://doi.org/10.1039/c2nr31418e

[12] S. Zhan, M. Yu, J. Lv, L. Wang, P. Zhou, Colorimetric Detection of Trace Arsenic(III) in Aqueous Solution Using Arsenic Aptamer and Gold Nanoparticles, Aust. J. Chem. 67 (2014) 813–818. https://doi.org/10.1071/CH13512

[13] F. Divsar, K. Habibzadeh, S. Shariati, M. Shahriarinourc, Aptamer conjugated silver nanoparticles for the colorimetric detection of arsenic ions using response surface methodology, Anal. Methods. 7 (2015) 4568-4576. https://doi.org/10.1039/C4AY02914C

[14] L. Song, K. Mao, X. Zhou, J. Hu, A novel biosensor based on Au@Ag core–shell nanoparticles for SERS detection of arsenic (III), Talanta 146 (2016) 285-290. https://doi.org/10.1016/j.talanta.2015.08.052

[15] N. Le Thao Nguyen, C.Y. Park, J.P. Park, S.K. Kailasa and T.J. Park, Synergistic molecular assembly of an aptamer and surfactant on gold nanoparticles for the colorimetric detection of trace levels of As^{3+} ions in real samples, New J. Chem. 42 (2018) 11530-11538. https://doi.org/10.1039/C8NJ01097H

[16] K. Vega-Figueroa, J. Santillán, V. Ortiz-Gómez, EO Ortiz-Quiles, BA Quiñones-Colón, Castilla- DA Casadiego, J Almodóvar, MJ Bayro, JA Rodríguez-Martínez, E. Nicolau, Aptamer-Based Impedimetric Assay of Arsenite in Water: Interfacial Properties and Performance, ACS Omega 3 (2018) 437-1444. https://doi.org/10.1021/acsomega.7b01710

[17] K. Matsunaga, Y. Okuyama, R. Hirano, S. Okabe, M. Takahashi, H. Satoh, Development of a simple analytical method to determine arsenite using a DNA aptamer and gold nanoparticles, Chemosphere 224 (2019) 538-543. https://doi.org/10.1016/j.chemosphere.2019.02.182

[18] Spuches AM, Kruszyna HG, Rich AM, Wilcox DE. Thermodynamics of the As(III)-thiol interaction: arsenite and monomethylarsenite complexes with glutathione, dihydrolipoic acid, and other thiol ligands. Inorg Chem. 2005 Apr 18;44(8):2964-72. https://doi.org/10.1021/ic048694q

[19] Zong C, Zhang Z, Liu B, Liu J. Adsorption of Arsenite on Gold Nanoparticles Studied with DNA Oligonucleotide Probes. Langmuir. 2019 Jun 4;35(22):7304-7311. https://doi.org/10.1021/acs.langmuir.9b01161

[20] Zong C, Liu J. The Arsenic-Binding Aptamer Cannot Bind Arsenic: Critical Evaluation of Aptamer Selection and Binding. Anal Chem. 2019 Aug 20;91(16):10887-10893. https://doi.org/10.1021/acs.analchem.9b02789

Materials Research Proceedings **16** (2020) 56-64
https://doi.org/10.21741/9781644900710-7

Growth and Optical Characterisation of Lithium Fluoride Films for Proton Beam Detectors

Maria Aurora Vincenti[1 *], Mauro Leoncini[2], Stefano Libera[1], Enrico Nichelatti[3], Massimo Piccinini[1], Alessandro Ampollini[1], Luigi Picardi[1], Concetta Ronsivalle[1], Alessandro Rufoloni[1] and Rosa Maria Montereali[1]

[1]Fusion and Technologies for Nuclear Safety and Security Dept., ENEA C.R. Frascati, Frascati (Rome), Italy

[2]CNR NANOTEC, Campus Ecotekne, Lecce, Italy

[3]Fusion and Technologies for Nuclear Safety and Security Dept., ENEA C.R. Casaccia, S. Maria di Galeria (Rome), Italy

*aurora.vincenti@enea.it

Keywords: Lithium Fluoride, Thin Films, Colour Centres, Photoluminescence, Radiation Detectors

Abstract. Polycrystalline LiF thin films were grown by thermal evaporation on glass, fused silica (Suprasil®) and Si(100) substrates in controlled conditions. Starting from the measured specular reflectance and direct transmittance spectra, some physical parameters of the LiF films grown on fused silica substrate were determined by using a best-fit procedure. The LiF films grown on glass and Si(100) substrates were irradiated by proton beams of 27 MeV nominal energy produced by a pre-clinical linear accelerator at several doses in the range between 4.2×10^3 and 1.7×10^5 Gy. Substrate-enhanced photoluminescence intensity was observed in coloured LiF films grown on Si substrates with respect to LiF films deposited on glass in the same deposition run. This behaviour is mainly ascribed to the reflective properties of the Si substrate in the visible spectral range, where the absorption and emission bands of F_2 and F_3^+ CCs are located, although other complex effects due to the polycrystalline nature of the films cannot be excluded. Further systematic studies are under way.

Introduction

Optical and chemical-physical properties of lithium fluoride (LiF) and its sensitivity to ionising radiations make this material promising for several applications ranging from photonics [1] to dosimetry [2]. LiF crystals are hard and almost non-hygroscopic, two precious properties for applications. LiF band gap is greater than 14 eV, so it is optically transparent from 120 nm to 7 μm and for this reason it is a widely used window material, in particular in the UV spectral region. Irradiations by ionising radiations, such as X-rays, γ-rays, protons, neutrons, electrons, etc., induces the formation of electronic defects, known as colour centres (CCs), which are stable at room temperature (RT). Among them, the aggregate F_2 and F_3^+ CCs (two electrons bound to two and three close anion vacancies, respectively) have almost overlapped absorption bands peaked at wavelengths of 444 and 448 nm, respectively; these together form the M absorption band, located in the blue spectral region [3]. Under optical excitation in the M band, the F_2 and F_3^+ CCs simultaneously emit broad Stokes-shifted photoluminescence (PL) bands peaked at 678 and 541 nm, respectively.

In recent years, the area of growth and characterisation of LiF films for radiation imaging detectors has seen a growing interest. LiF-film-based detectors can be grown by thermal evaporation on different substrates in controlled conditions, tailoring the appropriate geometry, size and thickness. These detectors are multi-purpose, allowing to detect X-rays, protons, neutrons, electrons, etc.. They are insensitive to the ambient light and they do not need any development process after the exposure. Their reading technique is based on the optical detection of the PL signal emitted by radiation-induced F_2 and F_3^+ CCs. LiF films grown by thermal evaporation were proposed and tested as novel extreme-ultraviolet (20 < hν < 300 eV) and soft X-rays (0.3 < hν < 3 keV) imaging detectors [4,5] based on F_2 and F_3^+ PL, as well as nuclear sensors for neutrons [6], for the characterisation of coherent X-ray sources [7] and advanced diagnostics of proton beams [8,9].

In this work preliminary experimental results about the enhancement of the emission properties of LiF-film-based detectors irradiated by 27 MeV proton beams at doses in the range between 4.2×10^3 and 1.7×10^5 Gy are presented and discussed.

Materials and methods

Polycrystalline LiF films, of nominal thicknesses 660, 1160 and 1450 nm, were grown by thermal evaporation on glass, fused silica and Si(100) substrates at ENEA C.R. Frascati. The deposition processes were performed in a steel vacuum chamber at a pressure below 1 mPa. After substrate cleaning, performed by using detergents in ultrasonic baths, the substrates were mounted on a rotating sample-holder placed above a tantalum crucible at a distance of 22 cm from it. During the deposition process, the substrate temperature was kept constant at 300 °C by means of four infrared halogen lamps controlled by the WEST 6400 thermoregulator and a calibrated thermocouple positioned inside the vacuum chamber. LiF powder (Merck Suprapur, 99.99% pure), placed in the crucible, was heated at about 850 °C by Joule effect. The deposition rate and the film thickness were monitored in situ by an INFICON quartz oscillator. All the deposition processes were performed keeping the deposition rate at a value of 1 nm/s.

After the growth, the thickness of the deposited LiF films was measured by using a Tencor P-10 stylus profilometer.

The morphological analysis was performed by means of a PARK System Atomic Force Microscopy (AFM), model XE-150, operating in air in non-contact mode.

The LiF films were irradiated in air at RT by proton beams at nominal energy of 27 MeV, produced by the injector of the pre-clinical linear accelerator TOP-IMPLART (Oncological Therapy with Protons–Intensity Modulated Proton Linear Accelerator for RadioTherapy) under development at ENEA C.R. Frascati [10]. The irradiations were performed placing the LiF samples perpendicularly to the proton beam at a distance of 10 mm from the 50 μm thick kapton window, which constitutes the exit of the machine beamline. The average beam current was 27.4 μA with a pulse charge of 77 pC/pulse and a repetition rate of 10 Hz. The irradiations were performed at several doses in the range between 4.2×10^3 and 1.7×10^5 Gy. According to the simulations made with SRIM (The Stopping and Range of Ions in Matter) software [11], the linear energy transfer of protons can be considered as constant through the limited thickness of LiF films and equal to ~ 4.5 keV/μm. The 27 MeV proton range within LiF is about 3.5 mm, much larger than the film thicknesses, therefore only a very small fraction of the total proton energy is released within the films.

PL and photoluminescence-excitation (PLE) spectra were collected at RT with a Horiba Scientific Fluorolog-3 spectrofluorimeter Model FL3-11, equipped with a 450 W xenon lamp, automatic slits, single-grating excitation and emission spectrometers, and a Hamamatsu R928 photomultiplier, adopting a front-face detection geometry.

 Materials Research Forum LLC
https://doi.org/10.21741/9781644900710-7

Results and discussion

The main deposition parameters, such substrate material, substrate temperature (T_s), deposition rate, film thickness, together with the method of substrate cleaning before coating, influence the film structural, morphological and optical properties, and the adhesion of the film to the substrate. When the substrate temperature is T_s = 300 °C, the T_s/T_m ratio, where T_m is the melting temperature of the film material and both temperatures are expressed in Kelvin, is greater than 0.5, which corresponds to Zone III of the Structure Zone Model [12]. In these experimental conditions, the film growth is characterized by bulk diffusion of admolecules, with activation energies above 0.3 eV, resulting in a rough equiaxed grained real structure of the film. Figure 1 shows the 3D AFM images over an area of (5×5) μm^2 of two LiF films, of nominal thickness 660 nm, grown on glass and Si(100) substrates in the same deposition run. The measured values of the Root Mean Square (RMS) roughness are about 9.6 nm for the LiF film grown on glass substrate and 11.0 nm for the LiF film deposited on Si(100) substrate.

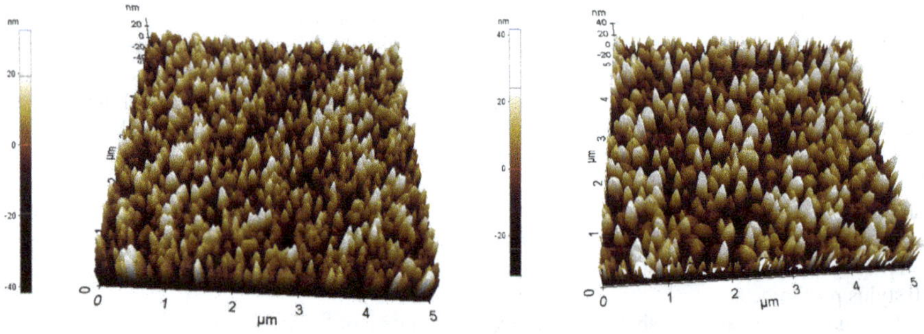

Fig. 1. 3D AFM images over an area of (5×5) μm^2 of two LiF films, of thickness ~660 nm, thermally-evaporated in the same deposition run on glass, left, and Si(100), right, substrates.

Optical and physical properties of the LiF films grown on Suprasil® were studied by analysing absolute specular reflectance and direct transmittance spectra measured in the 190÷1600 nm wavelength range by using a Perkin-Elmer Lambda 900 spectrophotometer. By best fitting the experimental spectra with the film model introduced in [13,14], several features of the layer could be quantitatively estimated. As an example, Fig. 2 shows the measured specular reflectance and direct transmittance spectra, *R* and *T*, of a LiF film of nominal thickness 660 nm, together with their best fitting curves. The *R* and *T* spectra of a bare silica substrate, twin of the one on which the LiF film was evaporated, were also measured and are shown in Fig. 2 too. From the best fit procedure an average thickness of (625 ± 5) nm, a surface root-mean-square roughness of (16.3 ± 0.2) nm, a deviation from parallelism of the film faces of (1.9 ± 0.3) % and a linear inhomogeneity along the growth axis of (-1.5 ± 0.4)% were obtained.

The analysis allowed deriving the spectral dispersions of the film material refractive index and extinction coefficient as well, which are plotted in Fig. 3.

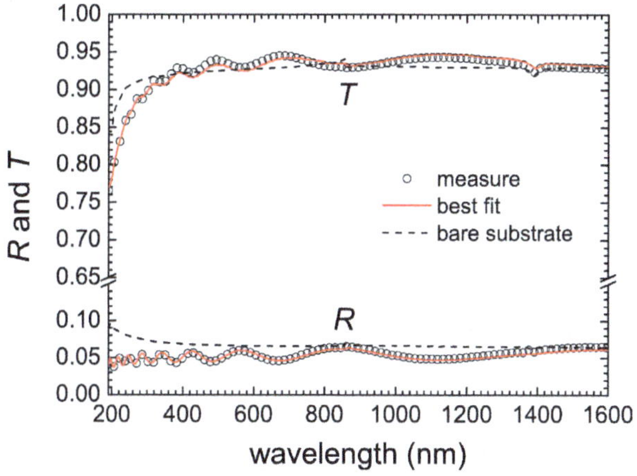

Fig. 2. Specular reflectance, R, and direct transmittance, T, spectra of the examined LiF film on silica substrate as measured at a Perkin-Elmer Lambda 900 spectrophotometer and corresponding best-fitting theoretical spectra. The measured reflectance and transmittance of the bare silica substrate are also shown.

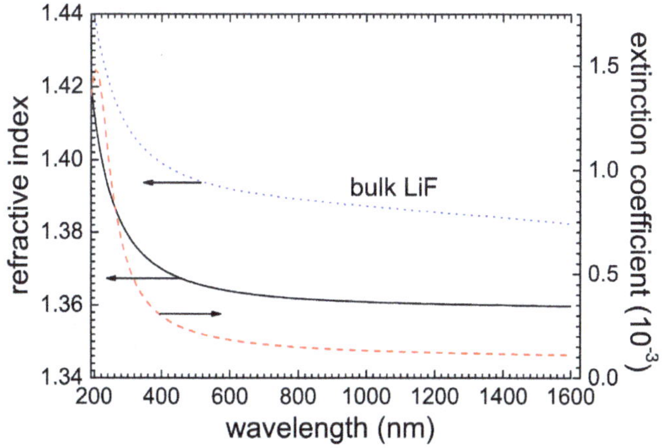

Fig. 3. Spectral dispersions of the LiF film refractive index (left scale) and extinction coefficient (right scale). The refractive index of bulk LiF [15] is also shown for comparison.

Materials Research Forum LLC
https://doi.org/10.21741/9781644900710-7

By comparing the refractive-index dispersion with tabulated spectral data of bulk LiF [15], also plotted in Fig. 3, another important feature of the film material was estimated: the packing density. This quantity represents the fraction of volume in the layer which is effectively occupied by LiF, the remaining fraction very likely consisting of air voids. Under the assumption that the sizes of such inhomogeneities are much smaller than the wavelengths of the incoming radiation, so that they are not sensed by it, this LiF-voids mixture can be approximately considered as a homogenous material whose refractive index can be evaluated by applying a suitable effective-medium model [16]. In our case, application of the Maxwell Garnett equation [16] allowed finding the packing density value that gives the best superposition of the refractive-index spectral dispersion of a bulk-LiF-voids mixture with that obtained for the examined film; the result was an estimated packing density of about 93%. Directly related to the packing density is the mass density of the film material: as a matter of fact, the latter should reasonably be equal to a weighted average of the mass densities of bulk LiF and air, the corresponding weights being equal to 93% and 7%, respectively, for the examined sample. Assuming a bulk LiF mass density of 2.635 g/cm^3 [17], a mass density of about 2.451 g/cm^3 was estimated for the LiF layer.

LiF films irradiated by 27 MeV proton beams get uniformly coloured along their thickness [8]. Figure 4 (left) shows the PL spectra of the LiF films, of the three different thicknesses (~ 660, 1165 and 1450 nm), grown on glass and Si(100) substrates in the same deposition run, irradiated by 27 MeV proton beams at the highest dose of 1.7×10^5 Gy, optically pumped at a wavelength of 445 nm, which simultaneously excite the PL of F_2 and F_3^+ CCs. Each PL spectrum consists of the superposition of two broad emission bands due to these two radiation-induced defects, peaked at the wavelengths of 678 and 541 nm, respectively. For each PL spectrum, the partial contributions of F_2 and F_3^+ CCs can be decoupled by using a best fit procedure with two Gaussian bands with fixed peak positions and widths, in agreement with literature available data [18]. Figure 4 (right) reports, as an example, the PL spectrum of the LiF film grown on glass substrate, thickness 1165 nm, together with such a best fit; the partial Gaussian band contributions due to F_2 and F_3^+ defects are also reported.

The PL signal intensity, which depends on the number of radiation-induced CCs stored in LiF, increases with the film thickness both for LiF films grown on glass and Si(100) substrates. It also increases with the irradiation dose [8,9], although this behaviour is not discussed here. For LiF films grown on glass and Si(100) substrates in the same deposition run, the PL signal of the LiF grown on Si(100) results higher than that of the LiF film grown on glass, for all the three thicknesses. The main reason of the observed PL enhancement is the high reflectivity of silicon in the visible spectral range, where the absorption and emission bands of F_2 and F_3^+ CCs are located. The reflective substrate is able to redirect a large fraction of F_2 and F_3^+ PL towards the detection system, that otherwise, as in the case of glass substrate, gets almost completely lost. Moreover, this reflected signal can constructively interfere with the straightly emitted one for all the orders of multireflection and at all the emission angles for suitable ratios of the film thickness to the wavelength. Indeed, the planar structure constituted by the coloured LiF film over the reflective substrate is an elementary half microcavity [19, 20].

PLE measurements are an effective method to investigate the absorption spectral features of F_2 and F_3^+ defects stored in LiF films grown on glass and Si substrates. In the case of LiF films grown on glass substrate, optical absorption spectra do not allow to identify the presence of radiation-induced F_2 and F_3^+ CCs because of the interference fringes due to the non-negligible refractive index difference between film and substrate. For LiF films grown on Si, moreover, this investigation is precluded by the substrate opacity.

 Materials Research Forum LLC
https://doi.org/10.21741/9781644900710-7

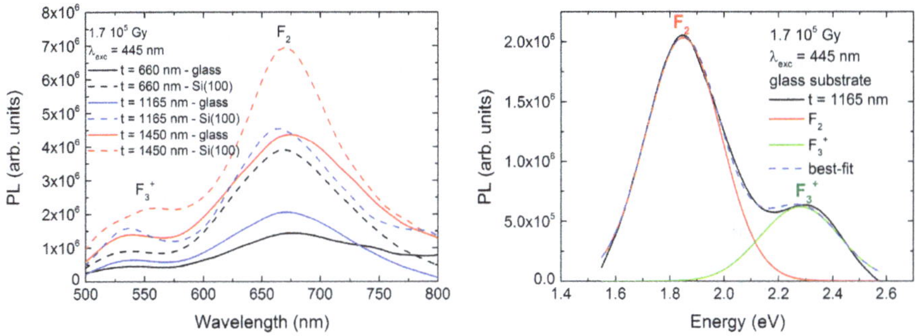

Fig. 4. (left) PL spectra of the LiF films thermally-evaporated on glass and on Si(100) substrates, of three thicknesses from 660 to 1450 nm, irradiated by 27 MeV proton beams at the dose of 1.7×10^5 Gy, measured at the excitation wavelength of 445 nm. (right) PL spectrum (solid black line) and best-fit (dashed blue line) of the LiF film grown on glass substrate, thickness ~1165 nm, irradiated by a 27 MeV proton beam at a dose of 1.7×10^5 Gy. The partial Gaussian band contributions due to F_2 (solid red line) and F_3^+ (solid green line) CCs obtained by best-fit, are also reported.

Figure 5 shows the PLE spectra of the LiF films grown on glass and Si substrates, nominal thickness 660 nm, irradiated at 1.7 10^5 Gy, collected at the emission wavelength of 541 and 678 nm. The blue vertical lines indicate the wavelength used to excite the PL spectra shown in Fig. 4.

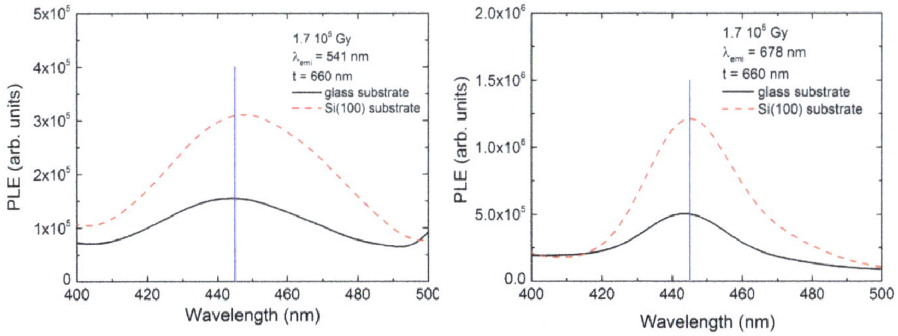

Fig. 5. (left) PLE spectra of the LiF films thermally-evaporated on glass and on Si(100) substrates, thickness ~ 660 nm, irradiated by 27 MeV proton beams at a dose of 1.7×10^5 Gy, acquired at the emission wavelength of 541 nm (contribution of F_3^+ CCs), left, and 678 nm (contribution of F_2 CCs), right. The vertical blue line indicates the wavelength used to simultaneously excite the PL spectra of the samples shown in Fig. 4.

For both LiF films grown on glass and Si(100) substrate, the F_3^+ PLE band is broader than the F_2 one and its peak position is located at longer wavelength, in agreement with the absorption spectral features of these radiation-induced defects known from literature [18].

The PLE signal collected at both the emission wavelengths is higher for the LiF film grown on Si substrate than for the LiF film grown on glass. An enhancement up to ~100% is observed for the F_2 PLE band peaks in the investigated dose range, in agreement with previous investigations performed at a lower proton energy [21]. This value is in agreement with the results of the model applied to Si-LiF film bilayers [20], which estimate an enhancement between 90% and 150%, as already observed in the PL spectra of Fig.4a. Again the main reason of this signal amplification is ascribed to the high reflectivity of silicon in the visible spectral range, where the absorption and emission bands of F_2 and F_3^+ CCs are located, although other complex effects due the presence of voids and grain boundaries in the polycrystalline matrix of the LiF films cannot be excluded.

Conclusions

Polycrystalline LiF thin films were grown by thermal evaporation on glass, fused silica(Suprasil®) and Si(100) substrates in controlled conditions. Some physical parameters of the LiF films grown on fused silica substrate were determined starting from the measured specular reflectance and direct transmittance films spectra by using a best-fit procedure.

LiF films grown on glass and Si(100) substrates were irradiated by proton beams of nominal energy 27 MeV at several doses in the range between 4.2×10^3 and 1.7×10^5. The investigation of PL and PLE spectra of radiation-induced F_2 and F_3^+ electronic defects shows a substrate-enhancement in coloured LiF films grown on Si substrates with respect to LiF films deposited on glass in the same deposition run, at the investigated doses. This behaviour is mainly ascribed to the reflective properties of the Si substrate in the visible spectral range, where the absorption and emission bands of F_2 and F_3^+ CCs are located, although other complex effects due to the polycrystalline nature of the films cannot be excluded.

Further systematic studies are under way, in order to investigate the influence of the polycrystalline nature of thermally-evaporated LiF films on the formation efficiency of F_2 and F_3^+ electronic defects and to obtain higher substrate-enhanced photoluminescence intensities for improving the performance of the LiF film-based radiation detectors especially at lower doses.

Acknowledgements

This research was partly carried out within the TOP-IMPLART (Oncological Therapy with Protons - Intensity Modulated Proton Linear Accelerator for RadioTherapy) project, funded by Regione Lazio, Italy.

References

[1] R.M. Montereali, Point defects in thin insulating films of lithium fluoride for optical microsystems, in: Handbook of Thin Film Materials, Vol. 3 (Academic Press) 2002, pp. 399-431. https://doi.org/10.1016/B978-012512908-4/50043-6

[2] S. W. S. McKeever, M. Moscovitch, P.D. Townsend, Thermoluminescence Dosimetry Materials: properties and uses, Nuclear Technology Publishing, Ashford, Kent TN23 1YW, England 1995.

[3] J. Nahum, Optical Properties and Mechanism of Formation of Some F-aggregate enters in LiF, Phys. Rev. 158 (1967) 814. https://doi.org/10.1103/PhysRev.158.814

[4] G. Baldacchini, S. Bollanti, F. Bonfigli, F. Flora, P. Di Lazzaro, A. Lai, T. Marolo, R.M. Montereali, D. Murra, A. Faenov, T. Pikuz, E. Nichelatti, G. Tomassetti, A. Reale, L. Reale, A. Ritucci, T. Limongi, L. Palladino, M. Francucci, S. Martellucci, G. Petrocelli, A Novel Soft X-Ray Submicron Imaging Detector Based on Point Defects in LiF, Rev. Scient. Instr. 76, (2005) 113104-1. https://doi.org/10.1063/1.2130930

[5] F. Cosset, A. Celerier, B. Barelaud, J.C. Vareille, Thin reactive LiF films for nuclear sensors, Thin Solid Films 303 (1997) 191-195. https://doi.org/10.1016/S0040-6090(97)00070-9

[6] G. Baldacchini, F. Bonfigli, A. Faenov, F. Flora, R.M. Montereali, A. Pace, T. Pikuz, L. Reale, Lithium Fluoride as a Novel X-Ray Image Detector for Biological μ-World Capture, J. Nanosci. Nanotechnol. 3 (6) (2003) 483-486. https://doi.org/10.1166/jnn.2003.023

[7] G. Tomassetti, A. Ritucci, A. Reale, L. Arizza, F. Flora, R.M. Montereali, A. Faenov, T. Pikuz, Two-Beam Interferometric Encoding of Photoluminescent Gratings in LiF Crystals by High-Brightness Tabletop Soft X-Ray Laser, Appl. Phys. Lett. 85 (18) (2004) 4163-4165. https://doi.org/10.1063/1.1812841

[8] M. Piccinini, F. Ambrosini, A. Ampollini, L. Picardi, C. Ronsivalle, F. Bonfigli, S. Libera, E. Nichelatti, M.A. Vincenti and R.M. Montereali, Photoluminescence of radiation-induced color centers in lithium fluoride thin films for advanced diagnostics of proton beams, Appl. Phys. Lett. 106 (2015) 261108. https://doi.org/10.1063/1.4923403

[9] M. Piccinini, E. Nichelatti, A. Ampollini, L. Picardi, C. Ronsivalle, F. Bonfigli, S. Libera, M.A. Vincenti and R.M. Montereali, Proton beam dose-mapping via color centers in LiF thin-film detectors by fluorescence microscopy, EPL 117 (2017) 3704. https://doi.org/10.1209/0295-5075/117/37004

[10] C. Ronsivalle, M. Carpanese, C. Marino, G. Messina, L. Picardi, S. Sandri, E. Basile, B. Caccia, D. M. Castelluccio, E. Cisbani, S. Frullani, F. Ghio, V. Macellari, M. Benassi, M. D'Andrea, and L. Strigari, The TOP-IMPLART Project, Eur. Phys. J. Plus 126 (2011) 7. https://doi.org/10.1140/epjp/i2011-11068-x

[11] J. F. Ziegler, M. D. Ziegler, J. P. Biersac, Optical Properties and Mechanism of Formation of Some F-aggregate centers in LiF, Nucl. Instrum. Methods *B* 268 (2010) 1818.

[12] P.B. Barna, M. Adamik, Fondamental structure forming phenomena of polycrystalline films and the structure zone model, Thin Solid Films 317 (1996) 27-33. https://doi.org/10.1016/S0040-6090(97)00503-8

[13] M. Montecchi, R.M. Montereali, E. Nichelatti, Reflectance and transmittance of a slightly inhomogeneous film bounded by rough, unparallel interfaces, Thin Solid Films 396 (2001) 262-273. https://doi.org/10.1016/S0040-6090(01)01253-6

[14] M. Montecchi, R.M. Montereali, E. Nichelatti, Erratum to Reflectance and transmittance of a slightly inhomogeneous film bounded by rough, unparallel interfaces, Thin Solid Films 396 (2001) 262-273. https://doi.org/10.1016/S0040-6090(01)01253-6

[15] E.D. Palik, Handbook of Optical Constants of Solids, Academic Press, San Diego, 1985.

[16] G.A. Niklasson, C.G. Granqvist, O. Hunderi, Effective medium models for the optical properties of inhomogeneous materials, Appl. Opt. 20 (1981) 26-30. https://doi.org/10.1364/AO.20.000026

[17] P. Patnaik, Handbook of Inorganic Chemicals, McGraw-Hill, New York, 2003.

[18] G. Baldacchini, E. De Nicola, R. M. Montereali, A. Scacco, V. Kalinov, Optical bands of F_2 and F_3^+ centers in LiF, J. Phys. Chem. Solids 61 (2000) 21. https://doi.org/10.1016/S0022-3697(99)00236-X

[19] R.M. Montereali, F. Bonfigli, M.A. Vincenti and E. Nichelatti, Versatile lithium fluoride thin-film solid-state detectors for nanoscale radiation imaging, Il Nuovo Cimento C 36 (2) (2013) 35.

[20] E. Nichelatti, R.M. Montereali, Photoluminescence from a homogeneous volume source within optical multilayer: analytical formulas, J. Opt. Soc. Am. A 29 (2012) 303. https://doi.org/10.1364/JOSAA.29.000303

[21] M. Leoncini, M.A. Vincenti, F. Bonfigli, S. Libera, E. Nichelatti, M. Piccinini, A. Ampollini, L. Picardi, C. Ronsivalle, A. Mancini, A. Rufoloni, R.M. Montereali, Optical investigation of radiation-induced color centers in lithium fluoride thin films for low-energy proton-beam detectors, Opt. Mat. 88 (2019) 580-585. https://doi.org/10.1016/j.optmat.2018.12.031

Keyword Index

Arsenic Detection 46

Carbon Materials 16
Chemical Oxidation 16
CMOS 1
Colour Centres 56
$CsPbBr_3$ Nanocrystals 27

EOMs 27

Glass Defects 38
Glasses 38

Heavy Metals 16
Hg(II) Ions Detection 6

Kerr Effect 27

Lithium Fluoride 56
Localized Surface Plasmon Resonance (L-SPR) 6

Mach-Zehnder 1
Metal Nanomaterials 6

NLO Materials 27

Oligonucleotide Aptamers 46
Optical Properties 38
Optical Sensors 6

Photoluminescence 16, 56
Plasma Treatment 38

Quenching 16

Radiation Detectors 56

Sensors 16
SERS 46
Silver Nanoparticles 6
Spectroscopy 16

Thin Films 56

VLSI 1

Z-Scan 27

About the editor

Paolo Prosposito, PhD, professor at the Industrial Engineering Department of the University of Rome Tor Vergata, Italy. Visiting professor at Shanghai University (China); Endeavour Fellowship of the Australian Government at School of Chemistry, Physics and Mechanical Engineering – Queensland University of Technology (Australia); Visiting scientist at Van 't Hoff Institute - University of Amsterdam (the Netherlands).

Main skills and expertise are on experimental Physics with special regard on Material Science and Optical Spectroscopy. Field of interest: organic vapour sensors based on luminescent quantum dots, synthesis and characterization of metallic nanoparticles for sensing applications, fabrication of 3D scaffolds for tissue engineering with photolithographic techniques, synthesis and characterization of photonic crystals, nanoimprinting and soft-lithographic techniques, surface nanostructuring of solar cells by soft lithography.

Author of more than 100 papers on peer reviewed scientific international journals and one patent.

www.ingramcontent.com/pod-product-compliance
Lightning Source LLC
Chambersburg PA
CBHW061050220326
41597CB00018BA/2731
9781644900703